Building Thinking Classrooms in Mathematics exudes enthusiasm for students, how they think, and how those thoughts coalesce into powerful thinking classrooms. It's also deeply practical, describing how everything from the teacher's questions to the arrangement of the furniture can add to your students' learning.

Dan Meyer
Chief Academic Officer, Desmos

If your students are not the ones doing the thinking in your classroom, then this book is for you! Peter Liljedahl provides concrete advice on each of 14 research-based practices, along with answers to frequently asked questions and suggestions for getting started, which will help you build a classroom where student thinking is the norm.

Peg Smith
Co-Author, *5 Practices for Orchestrating Productive Mathematics Discussions*
Professor Emerita, University of Pittsburgh
Pittsburg, PA

Peter Liljedahl's Thinking Classroom framework transformed my Mathematics classroom overnight. I was frustrated that despite my best teaching efforts some of my students still couldn't solve simple problems by their final exam. This framework gave me a starting point that I started implementing the very next day (don't wait for September to try this!) and next steps to continue incorporating as my practice evolved with 14 elements of the Thinking Classroom. Students began to talk to each other, think through complex problems, rely less on me and more on each other and best of all had better success in the courses I taught. The Thinking Classroom framework was exactly what my students and I needed!

Laura Wheeler
High School Math Teacher
Ottawa Carleton District School Board
Ontario, Canada

Peter refers to his research as "mucking about," and that is the key thing for me, that he goes into actual classrooms, and does math with students. We learn the most from being in actual classrooms, talking to students, and figuring out how they think about mathematics tasks. We need our students to be better thinkers, and to see mathematics for what it is: a beautiful way of thinking. We need them to see that they, too, can have powerful insights into interesting mathematics problems.

Matthew Oldridge
Author of *Teaching Mathematics Through Problem-Solving in K–12 Classrooms* (Rowman & Littlefield, 2018)
Teacher, Peel District School Board
Ontario, Canada

An in-depth action plan backed with significant research and data, Liljedahl's plan is one that can improve every classroom for the better, and he foresees and addresses any questions or concerns you may have regarding implementation. It is clear Liljedahl understands the students I teach in his list of student behaviors when posed with a now-you-try-one activity: the slackers, stallers, fakers, mimickers, and the few try-it-on-their-own-ers. This book outlines methods to increase the thinking and engagement of all my students. I was able to implement many of the methods the very next day.

Leslie Mohlman
Mathematics Teacher
Alpine School District
Lehi, UT

Peter Liljedahl's work is accessible, inspired by research, and embedded in classroom practice. He digs deeply and concisely into what it means to teach, learn, and assess in a thinking mathematics classroom. Elementary teachers, especially, will recognize themselves in this resource. Peter makes visible the often-intuitive moves of elementary classroom teachers, describing what it is we are doing when it all just works, and how to meaningfully shift our practice when it doesn't. From the way the furniture is arranged to how mathematical questions are posed, from who holds the pen to how to foster productive struggle and resilience, Peter sets the stage for genuine mathematical engagement in learners of all ages.

Carole Fullerton
Mathematics Teacher Leader and National Mathematics Consultant
Mind-Full Educational Consulting
Vancouver, BC

Research in education that turns right around and informs our practice is invaluable in today's schools and classrooms. Peter uses evidence gathered in mathematics classrooms to directly inform how we make changes to our teaching and learning that enhances learning. This is the essence of evidence-based practice, practice based on evidence from the very classrooms we seek to influence.

John Almarode
Associate Professor of Education
James Madison University
Harrisonburg, VA

After years of leading lessons in an "*I do, we do, you do*" format, I found that my students lacked a productive disposition toward mathematics and would give up on problems easily. I knew something had to change, but *what* was I going to change in my teaching practice and *how* was I going to get there? After 10 years of experimenting with different pedagogical approaches, classroom environment setups, and developing my own content knowledge, I realize that this book is the resource that could have helped me expedite the transformation I was after—moving from a classroom of "mimickers" to building a classroom of "thinkers." Save yourself years of experimentation by investing a few hours reading this excellent book. Your students will thank you.

Kyle Pearce
K–12 Mathematics Consultant
MakeMathMoments.com & Greater Essex County District School Board
Ontario, Canada

Building Thinking Classrooms is an instructional *tour de force* for any math teacher. From his extensive research, Peter offers remarkably actionable classroom structures and teacher facilitation moves that get students to think and move forward in their thinking. I'm thrilled it's finally here!

Fawn Nguyen
Math TOSA, Rio School District
Oxnard, CA

For years I have heard about Thinking Classrooms in workshops, articles, and online. This engaging book has taken all the pieces that I have heard and seen and presents them in an easy to read, and more importantly, actionable package. Things that seemed a little too "I could never do that" for me now seem doable and I am inspired to begin to make changes. I am left with plenty to reflect upon in my current practice even as I begin to think about moving to a Thinking Classroom.

Casey McCormick
Math Teacher, Grades 5–8
Citrus Heights, CA

Building Thinking Classrooms prompts us to reflect on the potential of mathematics classrooms, teachers, and learners. Supported by numerous stories from classrooms, Peter methodically exposes the familiar structures of school mathematics that suppress the potential of learners, then carefully outlines a set of opportunities around which teachers of mathematics can organize a dynamic and responsive classroom.

Nat Banting
Recipient of the 2019 Margaret Sinclair Memorial Award and
the National Museum of Mathematics' 2019 Rosenthal Prize
Mathematics Teacher
Saskatchewan, Canada

Though there are many innovations in the area of teaching mathematics, few speak with a particular lens in terms of setting up an environment where thinking is made visible, where it's public, where positive interdependence is connected to individual and/or group accountability, with students relying on their own agency, as well as the wisdom of their peers. One where the teacher is freed up to have eyes on all student work, and watch the thinking process in action. In other words, thinking becomes a clearly visible driver in this environment. All of this supports the release of responsibility to the students. It honors their voices, allows for the bumps in learning, and makes the thinking more public, thus supporting and encouraging risk-taking in a safe and supportive environment.

Yana Ioffe
School Principal
Corwin Consultant
Ontario, Canada

This book is timely and provides an accurate portrayal of what is occurring in mathematics classroom across the country. The book is a valuable reflexive tool that teachers can use as they analyze their own teaching practices.

Kenneth Davis
High School Mathematics Teacher and Department Chair
School District of Beloit
Beloit, WI

Peter's work in building thinking classrooms has been the single most impactful (driver for) change in secondary mathematics education that I have witnessed. I have never seen another idea/approach/model capture so many teachers immediately, and make it past the point from learning to actual implementation in almost every classroom or instance that I have witnessed.

Mishaal Surti
Educational Consultant
Ontario, Canada

For teachers hoping to transform their teaching practice, Peter has written a definitive source. Peter's conversational style makes this work both interesting to read and easy to follow. He describes a rich set of practices that will help mathematics teachers transform, in a positive way, everything about their classroom. Peter turns the daunting challenge into something manageable with advice that is both believable and practical.

David Wees
Senior Curriculum Designer
DreamBox Learning
British Columbia, Canada

BUILDING THINKING CLASSROOMS IN MATHEMATICS

GRADES K-12

For every teacher who has had the courage to change.

BUILDING THINKING CLASSROOMS IN MATHEMATICS

14 Teaching Practices for Enhancing Learning

GRADES K-12

Peter Liljedahl

Foreword by Tracy Johnston Zager
Illustrations by Laura Wheeler

CORWIN Mathematics

For information:

Corwin
A SAGE Company
2455 Teller Road
Thousand Oaks, California 91320
(800) 233–9936
www.corwin.com

SAGE Publications Ltd.
1 Oliver's Yard
55 City Road
London, EC1Y 1SP
United Kingdom

SAGE Publications India Pvt. Ltd.
B 1/I 1 Mohan Cooperative Industrial
   Area
Mathura Road, New Delhi 110 044
India

SAGE Publications Asia-Pacific Pte. Ltd.
18 Cross Street #10–10/11/12
China Square Central
Singapore 048423

Publisher, Mathematics: Erin Null
*Associate Content Development
   Editor:* Jessica Vidal
Editorial Assistant: Caroline Timmings
Production Editor: Tori Mirsadjadi
Copy Editor: Cate Huisman
Typesetter: Integra
Proofreader: Susan Schon
Indexer: Integra
Cover Designer: Gail Buschman
Illustrator: Laura Wheeler
Marketing Manager: Margaret O'Connor

Printed in the United States of America.

Library of Congress Cataloging-in-Publication Data
Names: Liljedahl, Peter, 1967- author.
Title: Building thinking classrooms in mathematics, grades K-12 : 14
   teaching practices for enhancing learning / Peter Liljedahl.
Description: Thousand Oaks, California : Corwin, 2021. | Includes
   bibliographical references.
Identifiers: LCCN 2020032528 | ISBN 9781544374833 (paperback) |
   ISBN 9781544374864 (adobe pdf) | ISBN 9781544374840
   (ebook) | ISBN 9781544374871 (ebook)
Subjects: LCSH: Mathematics--Study and teaching (Elementary) |
   Mathematics--Study and teaching (Secondary) | Effective teaching. |
   Classroom environment.
Classification: LCC QA135.6 .L55 2021 | DDC 510.71/2--dc23
LC record available at https://lccn.loc.gov/2020032528
This book is printed on acid-free paper.

20 21 22 23 24 10 9 8 7 6 5 4 3 2 1

CONTENTS

CHAPTER 1: WHAT TYPES OF TASKS WE USE IN A THINKING CLASSROOM   18

CHAPTER 2: HOW WE FORM COLLABORATIVE GROUPS IN A THINKING CLASSROOM   38

CHAPTER 3: WHERE STUDENTS WORK IN A THINKING CLASSROOM

CHAPTER 4: HOW WE ARRANGE THE FURNITURE IN A THINKING CLASSROOM

CHAPTER 5: HOW WE ANSWER QUESTIONS IN A THINKING CLASSROOM

CHAPTER 6: WHEN, WHERE, AND HOW TASKS ARE GIVEN IN A THINKING CLASSROOM

CHAPTER 7: WHAT HOMEWORK LOOKS LIKE IN A THINKING CLASSROOM

CHAPTER 8: HOW WE FOSTER STUDENT AUTONOMY IN A THINKING CLASSROOM

CHAPTER 9: HOW WE USE HINTS AND EXTENSIONS IN A THINKING CLASSROOM

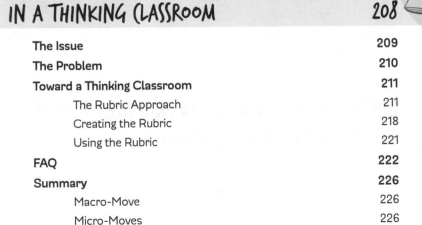

CHAPTER 13: HOW WE USE FORMATIVE ASSESSMENT IN A THINKING CLASSROOM   230

CHAPTER 14: HOW WE GRADE IN A THINKING CLASSROOM   252

CHAPTER 15: PULLING THE 14 PRACTICES TOGETHER TO BUILD A THINKING CLASSROOM   279

FOREWORD

"Research says …"

In education, this phrase is often followed by claims about what works in teaching and learning. I regularly find myself responding with clarifying questions:

"What do you mean by *research*? How was the study designed?"

"How was *success* defined? Success for whom, under what conditions?"

"What do you mean by *effective*? How was effectiveness *measured* in this research study you're citing?"

Sometimes, the answers are underwhelming. For example, I'm rarely convinced we should change teaching practices based on studies that used a single measure of success, such as annual standardized test scores, and that were conducted without researchers setting foot in the subject classrooms to see what was actually going on. Other times, I'm enthralled and informed by research, if that research involved thoughtful and serious study of teaching, learning, and student thinking.

Peter Liljedahl's work is that second kind of research, the research I can't get enough of. In my reading of Liljedahl's work, *research* means exploring important, testable questions with more than four hundred teachers and their thousands of students over 15 years. *Success* means getting more of these students thinking in math class, for longer amounts of time. *Effective* describes teaching decisions and practices that create conditions for student thinking. And results are *measured* by watching what students do: How many seconds does it take students to get to work? How long do they persevere? How engaged are they? How often do they pull out their phones for distraction? Who is participating? How much do students talk? How much does mathematical knowledge move from one group to another?

Liljedahl is a studier of students. By valuing, observing, and interviewing them, he has gathered incredibly useful information about one of the slipperiest questions in education: *What works?* For example, does it make a difference if teachers assign students a task by projecting it, printing it, writing down a textbook page number, or explaining it verbally? (Turns out, yes!) While teachers introduce this assignment, does it matter if students sit or stand? (It does.) How much of an impact does the timing of the launch within the lesson have? (A lot.) While students work on the task, should they use notebooks, chart paper, or erasable surfaces? In groups or individually? If they're in a group, should everyone have a marker, or just one student? What's the optimal group size? How should these groups be formed? How frequently should they change? These are just a few of the hundreds of questions Liljedahl and his colleagues considered and tested through tens of thousands of hours of classroom experimentation to figure out what works and what matters. When they discovered a technique that yielded a significant benefit during a two-week trial—as measured by increased student engagement and thinking—they collaborated with teachers to refine the technique over several weeks, and then tested the results with many additional teachers in a wide range of settings over longer stretches of time. Only lasting techniques that produced the most student thinking and were transferable across teachers and schools have made it into this book.

The practical, readable resource you're holding in your hands is an enormous gift and guide to math teachers. Liljedahl has identified the most effective changes we can make to get our students thinking and keep them thinking longer. He has organized these shifts into an intuitive framework so we can start as soon as we are ready and tackle one piece at a time. Liljedahl uses common sense, refreshingly honest student voices, and everyday language to describe ideas and behaviors: when he talks about "now-you-try-one" tasks or "I-write-you-write" notes, we know just what he means. Every moment is grounded in classroom reality and the path ahead feels so doable because of the way he has laid it out for us. I found myself nodding regularly as I read, and I am grateful that he has organized and described his findings with such clarity that we can implement these shifts incrementally, in the most effective sequence.

Incremental doesn't mean gradual, however. Even though the specific shifts are practical and manageable, they will feel disruptive—that's actually the main idea. Liljedahl argues convincingly that we need to interrupt the entrenched patterns of school. Students arrive at our classroom doorsteps each year, week, and day expecting the same familiar script: They take their seats and we stand at the front of the room. We show them what to do on the board and they copy it down in their notes. They then mimic us through worked examples, again on homework problems, and finally on a test scored by percentage. Over the course of *Building Thinking Classrooms in Mathematics*, we learn from Liljedahl's contrarian logic to question and replace each of these familiar patterns with different, more effective, field-tested techniques. Instead of sitting during a discussion, students stand. Instead of taking mindless notes to please us, they take notes that would be helpful to their future forgetful selves. Instead of mimicking our methods alone, they think about new problems together, and so on. We make these changes not for the sake of change, nor for ideological reasons, but because these practices lead to increased student thinking in hundreds of diverse classrooms. Taken together, these practices signal to students that this class is different: In this class, they'll be expected to think.

Why does it matter? Because most of our students do an awful lot of "studenting," but not much thinking. Students from communities that have historically been excluded from mathematics are often denied access to thinking at all. For the health of our students and our societies, we need to challenge institutional norms and build thinking classrooms in which we value students' thinking and time rather than use legacy practices that encourage students to slack, stall, mimic, and fake their way through the system. In Chapter 9, Liljedahl wrote that the "goal of building thinking classrooms is not to find engaging tasks for students to think about. The goal of thinking classrooms is to build engaged students that are willing to think about any task."

Given the enormity of the problems we all face, I am especially eager for teachers to implement the ideas and techniques in *Building Thinking Classrooms*. Could there be anything more important and pressing than teaching students how to think?

—**Tracy Johnston Zager**
Author of *Becoming the Math Teacher*
You Wish You'd Had: Ideas and Strategies from Vibrant Classrooms

ACKNOWLEDGMENTS

It takes a tremendous number of people to produce a book. For *Building Thinking Classrooms in Mathematics*, it began with the hundreds of teachers who partook in the research. Without their willingness to change and their courage to let me into their classrooms, this book would not have been possible. I thank you for your professionalism and dedication to your craft. I also owe a debt of gratitude to teachers who have championed the Building Thinking Classrooms Framework among their peers. Your enthusiasm has been infectious and has contributed significantly to the success of the framework. I thank you for being such good ambassadors of this work. I want to thank my wife, Theresa. Whenever I was offered the possibility to work with one more teacher, one more school, or one more school division; or to fly off to give one more presentation on building thinking classrooms, her response was always the same: "If this is going to change the experience for even one student, you have to say yes." Your steadfast encouragement, support, and faith that I could make a difference made all this possible. Finally, I want to acknowledge the entire team at Corwin beginning with my editor and Corwin mathematics publisher, Erin Null. You not only had faith that *Building Thinking Classrooms* could be turned into a book, but you helped me find the voice necessary to do so. I want to thank Laura Wheeler for illustrating this book. Ever since I began writing about and presenting on building thinking classrooms, you have found ways to give life to my words through your artistry. And I thank content development editor Jessica Vidal, who worked side by side with Laura and me to imagine how this book would come to life and look and feel; editorial assistant Caroline Timmings and production editor Tori Mirsadjadi, who helped with many technical aspects of turning a manuscript into an actual book; and senior marketing manager Margaret O'Connor, who worked so hard to get this book into your hands.

Portions of the research presented in this book were funded through the Social Sciences and Humanities Research Council of Canada.

PUBLISHER'S ACKNOWLEDGMENTS

Corwin gratefully acknowledges the contributions of the following reviewers:

John Almarode
Associate Professor of Education
James Madison University
Harrisonburg, VA

Avital (Tali) Amr
Math Consultant, Curriculum and Instructional Services
York Region District School Board
Ontario, Canada

Kenneth Davis
High School Mathematics Teacher and Department Chair
School District of Beloit
Beloit, WI

Yana Ioffe
School Principal
Corwin Consultant
Ontario, Canada

Tricia Loney
Secondary Mathematics Learning Coordinator
Thames Valley District School Board
Ontario, Canada

Kristen Mangus
Mathematics Support Teacher
Howard County Public School System
Crownsville, MD

Casey McCormick
Math Teacher, Grades 5–8
Citrus Heights, CA

Ruth Harbin Miles
University Instructor and Retired K–12 Mathematics Supervisor
Mary Baldwin University
Madison, VA

Leslie Mohlman
Mathematics Teacher
Alpine School District
Lehi, UT

Fawn Nguyen
Math TOSA (Teacher on Special Assignment)
Rio School District
Oak View, CA

Matthew Oldridge
Teacher, author of *Teaching Mathematics Through Problem-Solving
in K–12 Classrooms* (Rowman & Littlefield, 2018)
Peel District School Board
Ontario, Canada

Kyle Pearce
K–12 Mathematics Consultant
MakeMathMoments.com & Greater Essex County District School Board
Ontario, Canada

Mishaal Surti
Educational Consultant
Ontario, Canada

David Wees
Senior Curriculum Designer
DreamBox Learning
British Columbia, Canada

Laura Wheeler
High School Math Teacher
Ottawa Carleton District School Board
Ontario, Canada

ABOUT THE AUTHOR

Photo courtesy of Simon Fraser University.

Dr. Peter Liljedahl is a professor of mathematics education at Simon Fraser University in Vancouver, Canada. He is the current president of the Canadian Mathematics Education Study Group (CMESG), past-president of the International Group for the Psychology of Mathematics Education (IGPME), editor of the *International Journal of Science and Mathematics Education* (IJSME), on the editorial board of five major international journals, and a member of the NCTM Research Committee. Peter has authored or coauthored numerous books, book chapters, and journal articles on topics central to the teaching and learning of mathematics. He is a former high school mathematics teacher who has kept his research close to the classroom and consults regularly with teachers, schools, school districts, ministries of education, and universities on issues of teaching and learning, assessment, and numeracy. Peter is a sought-after presenter who has given talks all over the world on the topic of building thinking classrooms, for which he has won the Cmolik Prize for the Enhancement of Public Education and the Fields Institute's Margaret Sinclair Memorial Award for Innovation and Excellence in Mathematics Education.

INTRODUCTION

When I first met Jane in 2003, she had been teaching middle school for 15 years. Although she was comfortable teaching math, there was a new curriculum on the horizon—and word on the street was that this new curriculum was going to have a heavier focus not only on problem solving, but also teaching *through* problem solving. In her 15 years of teaching, Jane had never done either of these. So, she decided she should get out in front of the new curriculum, learn something about problem solving, and start playing with it in her classroom.

Jane knew three things about me. First, she knew that I liked problem solving. My research at the time was, in essence, on creativity in problem solving, and I had been doing some workshops for teachers in her school district on this topic. Second, Jane knew that I was working on my PhD, was out of the classroom, and therefore had nothing but spare time on my hands. And third, she knew my e-mail address. I don't know how Jane knew any of these things, as I had never met, or even heard of, Jane. Nonetheless, one day in 2003 I received an e-mail from Jane:

> **Jane**   Hi. I'm interested in implementing problem solving in my Grade 7/8 mathematics classroom. Can I get some help from you?

Fantastic! I had been out of the classroom for a few years and I was missing teaching. To me this was an opportunity to not only get back into the classroom, but also do some problem solving with students.

> **Peter**   I'd love to help. Why don't we have a meeting to discuss it? I can come to school tomorrow. What room are you in and what time does school end?

So, the next day I showed up at Jane's door at 3:15 with a big smile on my face. This was going to be awesome.

Jane, who had clearly worked with researchers before, was not as enthusiastic.

> **Jane** Look. Before we start talking about problem solving, I want to get a few things straight. First, I don't want any of your glee and enthusiasm in here. I don't want to coteach with you. I don't even want to coplan with you. All I really wanted were some good problems that I could use in my Grade 7/8 math classroom. I don't even know why we are having this meeting.

This was not what I had been expecting. In fact, it was about as far from what I had been expecting as possible. But I would not be deterred, and after 15 minutes of discussion we arrived at a tense agreement—of sorts. I would give Jane *good* problems to try, and she, in return, would allow me to watch her implement them. But she had rules.

> **Jane** First, you have to stay in that desk [pointing at a desk in the back corner of the room]. You are not allowed to talk to the students. And you are definitely not allowed to talk to me.

And so it was that we began our collaboration—of sorts.

The first problem I gave Jane came from Lewis Carroll and was a problem I had used many times with my Grade 8s and 9s. I knew that this was a good problem. The context was engaging, the answer was non-trivial, and it didn't require any sophisticated mathematics to solve. And my students, when I had used it with them, had enjoyed arguing over the various answers they arrived at.

> **If 6 cats can kill 6 rats in 6 minutes, how many will be needed to kill 100 rats in 50 minutes? (Lewis Carroll, 1880)**

So, the next morning I sat in Jane's class and watched her write this problem up on the board for her students to solve. Before I tell you what happened next, let me review a few details. As mentioned, Jane had been teaching for 15 years and until this day had never used problem solving in her classroom. Her students sat in desks that were in rows with some of the rows put together to make student pairs (see Figure i.1). The students did not have assigned seats and sat and worked with who they wanted. A typical lesson, Jane had told me, began with her going over homework. This was followed by a lecture, during which time Jane demonstrated how to answer questions and the students took notes. Toward the end of the lesson Jane would ask students to do what I call *now-you-try-one* questions, which, after a few minutes, she would then go over. After a few of these she would assign homework out of the textbook, a student workbook, or a handout, and the students would work on this for the rest of the class. In short—it was a typical math class and a typical math lesson. Oh, and it was May—six weeks before the end of the school year.

Figure i.1 Students in a traditional classroom work on a task.
Source: skynesher/iStock.com

With that information in hand, how do you think her first attempt at using a problem-solving task like this with her students went? Yup—it was a disaster. As soon as Jane asked the students to solve the question on the board, a forest of hands went up and Jane started moving. She was going from student to student, from pair to pair, helping students who had questions about what they were supposed to do, if they were doing it right, and if this was the correct answer. Rather quickly, students became discouraged and began giving up, and now Jane was spending as much time encouraging students to keep going as she was helping the students who were still working.

Meanwhile, I was sitting in the back of the room, in my designated desk—not talking to the students and definitely not talking to Jane. The whole time I was watching this train wreck I was thinking that this was it—Jane was going to throw me out of her class, and that would be it for our brief, but spectacularly miserable, collaboration.

After about 25 minutes, Jane shifted gears and got the students onto a different activity, and she came up to me and said, "Give me another one." I was both shocked and impressed. There was more to Jane than met the eye. So, I gave Jane a second task, and the next morning I was back in my desk watching Jane try it again—same students, new problem.

It went worse. The students were quicker to give up, and Jane now spent more time encouraging and less time helping. At the end of the activity Jane came up to me and said, "Give me another one." This woman had grit. Over the last 18 years I have worked with hundreds of teachers, and not since Jane have I encountered a teacher with such fortitude—such will and determination to keep going in the face of utter failure. So, I gave Jane a third task, and the next morning I was again back in my desk—same students, new problem.

It was the worst of all. The students had absolutely no fight left in them, and for 25 minutes they just sat there, off task, and talking amongst themselves. Jane still had fight in her, however. And for the entirety of the 25 minutes she kept moving around the room trying to get something happening. When she came up to me at the end of the activity, she said, "I think we're done."

I agreed. Everybody in the room was in pain. The students were frustrated. Jane was exhausted. And I was disappointed. It was time to stop. But I wanted to understand why the tasks that I had used with success previously were failing so badly. So, I asked Jane if I could stay for the rest of day and watch her teach. She agreed and added, "You know the rules."

As it turns out, I sat in Jane's room for three full days watching her teach using her aforementioned routine of going over homework, demonstration, notes, now-you-try-one tasks, and assigning homework. Sometimes she was teaching the same students with whom she had tried the problem-solving tasks. Sometimes she taught other students. Toward the end of the third day, I was struck by two epiphanies. The first was the realization that at no point in the three days of observation had I seen Jane's students do any thinking—at least not the kind of thinking that we know students need to do to continue to be successful in mathematics in future grades. This is not to say that there was no activity. There was lots of activity—the students were busy from the beginning of class to the end. They were taking notes, answering questions, filling in worksheets, and starting on their homework. They were busy. They just weren't thinking.

The second epiphany was the sudden realization that Jane was planning her teaching on the assumption that students either couldn't or wouldn't think. Jane was in a tough position—she had a room full of students who weren't thinking, yet she had curriculum to get through and standards to meet. This is not uncommon. Every day, teachers all over the world find themselves in this exact same dilemma. Even teachers who, by traditional measures, are considered good teachers—who know their content, care about their students, and want to do the best for them—face this dilemma. Jane was considered, in her school and throughout her district, to be a very good teacher—her students performed well on tests, and no students appeared to be falling through the cracks. Jane wanted to do her best for her students, and she was willing to work hard to get there. And yet Jane found herself in this exact dilemma. So, what did she do? She did what many of us do—she structured activities that allowed her to move through the content as quickly and efficiently as possible without requiring her students to think. I'll give you an example.

There was an activity I watched Jane do that can be loosely described as a toothpick problem. The goal of the activity was to have students construct a row of squares out of toothpicks and record how many toothpicks it took to construct rows of different lengths. From these data, students were to then extrapolate and figure out how many it would take to build a row of length 10, 20, and 100 and then express the generalization in some prealgebraic format. These are great thinking activities when students are left to explore. In Jane's class, however, this activity was a set of instructions on a worksheet that she got from one of her resources. This wonderful patterning, extrapolation, and generalization activity had been reduced to a form of cookbook mathematics that ensured that, within 20 minutes or so, every student had completed it while, at the same time, ensuring that no one would do any thinking. Of course, these activities enabled the students to not have to think, which, in turn, forced Jane to keep planning her teaching on the assumption that students either couldn't or wouldn't think. But what choices did she have? Jane was stuck in a sort of endless and vicious non-thinking cycle. This is a problem. Thinking is a necessary precursor to learning, and if students are not thinking, they are not learning.

> Thinking is a necessary precursor to learning, and if students are not thinking, they are not learning.

I wondered if this was a uniquely Jane problem, so I visited another teacher in her school. I saw the same thing. I visited another—same thing. In all, I visited five teachers in that building, and everywhere I went I saw the same thing—students not thinking and teachers planning their teaching on the assumption that students either couldn't or wouldn't think. This is now a school problem.

I now wanted to see if this was a uniquely school problem, so I reached out to educators I knew and asked them to recommend to me teachers that they had heard were good. I contacted these teachers and asked if I could come in and watch them teach and watch their students learn. Many of them said yes. So, I left Jane's school and I visited different classrooms in different schools. When I was in those classrooms observing, I would ask those teachers if they knew of a teacher, in a different building, that they had heard was good. And so it was that I hopped from classroom to classroom, from school to school, visiting these good teachers.

Because I was following this thread of good teachers there was a lot of diversity among the schools I visited. I visited classrooms of every grade from kindergarten to Grade 12. I was in low socioeconomic settings and high socioeconomic settings. I was in French-speaking classrooms and English-speaking classrooms. I was in public schools and private schools. In all, I was in 40 different classrooms in 40 different schools. And everywhere I went I saw the same thing—students not thinking and teachers planning their teaching on the assumption that students either couldn't or wouldn't think. And, like Jane, these were all considered good teachers—they knew their content, they cared about their students, and they cared that their students got through the content. And, like Jane, these 40 teachers were all caught in the same sort of endless and vicious non-thinking cycle—they had students who weren't thinking, and they had content to get through. And, like Jane, they were using resources and textbooks that were designed to facilitate this. This is not a Jane problem. Or a Jane's school problem. This is a systemic problem (see Figure i.2).

> Everywhere I went I saw the same thing—students not thinking and teachers planning their teaching on the assumption that students either couldn't or wouldn't think.

Figure i.2 Students not thinking.
Sources: Goldfaery/iStock.com and Courtney Hale/iStock.com

STUDENTS *NOT THINKING*

At this point you may be satisfied with my statement that students were not thinking, and you may be nodding with the realization that that is also happening in your classroom, and you may be keen to get on with the rest of the book about how to change that—how to build a thinking classroom. If that is the case, then you can skip to the next section on institutional norms. If, however, you want a bit more of a description of what I mean by not thinking and how much of this was really happening in these 40 classrooms, then read on.

When I was visiting these 40 classrooms and coming to the realization that everywhere I went I saw students not thinking, what I really had was a *sense* that students were not thinking. I didn't have a good way to either qualify or quantify what I was seeing and not seeing. It was only a sense. It turned out to be true, but at the time it was only a sense.

My first effort to more precisely describe what I was seeing came later through a series of research projects into *studenting* behavior. Studenting, a term first coined by Fenstermacher (1986), is the analogue to teaching. As teachers, we do a great number of things that may or may not have to do with the facilitation of student learning. We take attendance, deal with classroom disruptions, make school announcements, collect permission forms, fund raise, and, oh yeah, we also help students learn the curricular content and develop some skills. All of these activities fall under the umbrella term of teaching. For Fenstermacher, studenting is the analogue to this.

> . . . there is much more to studenting than learning how to learn. In the school setting, studenting includes getting along with one's teachers, coping with one's peers, dealing with one's parents about being a student, and handling the non-academic aspects of school life. (1986, p. 39)

> [as well as] 'psyching out' teachers, figuring out how to get certain grades, 'beating the system,' dealing with boredom so that it is not obvious to teachers, negotiating the best deals on reading and writing assignments, threading the right line between curricular and extra-curricular activities, and determining what is likely to be on the test and what is not. (1994, p. 1)

Studenting: is what students do in a learning setting—some of which is learning.

In essence, studenting is what students do in a learning setting—some of which is learning. And much of which is not. For me, studenting was the perfect way to start thinking about what it is that students are doing if they are not thinking. So, I decided to begin

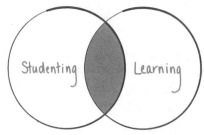

to research studenting within a number of what are called *activity settings* within the mathematics classroom. An activity setting is a discrete and well-defined activity within a lesson. The activity settings I first researched were now-you-try-one tasks, note-taking, and homework. I will present the results from note-taking and homework in Chapters 7 and 11, respectively. Here, I will present the results from the studenting research into now-you-try-one tasks.

A now-you-try-one task is a task that teachers ask students to do after the teacher has demonstrated to students how to do something. So, for example, we may be demonstrating to students how to multiply two-digit numbers, and after we have thoroughly explained this and done two or three examples, we may turn to our students and say, "Now you try one," as we write up the one we want them to try. And then we wait for 4 minutes and 22 seconds, which is the average amount of time teachers give students to do a now-you-try-one task, before we go over how to solve it. Then, in many cases, we give the students another now-you-try-one task. In my visits to the aforementioned 40 classrooms, now-you-try-one tasks were a foundational and central part of every lesson I observed and, for many of these teachers, were part of the fabric of what it means to teach.

When I asked these teachers to tell me what student behavior they expect to see during these moments, the answer was always the same.

Lillian	*I expect to see my students try it on their own.*
Researcher	*For what purpose?*
Lillian	*To see if they can do it, and to learn from their mistakes if they can't.*

We expect students to try it—and learn from it. Now-you-try-one tasks are a type of self-assessment where students and teachers learn whether the demonstrations were a success. This is pretty straightforward. So, what do students really do? What are their studenting behaviors during this discrete and well-defined learning setting? Well, it turns out that some students behave exactly as we expect—but only about 20% of them. The rest do not. In a study into studenting behaviors across several different classrooms, we found an array of behaviors[1] during the now-you-try-one activity setting (Liljedahl & Allan, 2013b). See if you recognize some of these.

---

[1] For a deep analysis of the psychology behind these, and other, studenting behaviors, see Allan (2017).

1. **Slacking** - A number of students in each class did not attempt the task at all. Instead, they spent the time looking at their smart phones, talking to other slackers, or literally doing nothing. When they were interviewed, it became clear that the students who slacked either didn't know what was going on or didn't care what was going on.

2. **Stalling** - Like the students who slacked, these students did not attempt the task. Unlike the slackers, however, these students filled the time with legitimate off-task behaviors like sharpening a pencil, getting a drink of water, going to the bathroom, or endlessly rooting in their backpack for some vital piece of equipment. When interviewed, these students told us that they either didn't know how to do the question or knew that if they just waited for a few minutes the teacher would go over it.

3. **Faking** - Some students pretended to do the task but were, in reality, doing nothing. Faking involved studiously looking at the board, flipping pages in the textbook, appearing to ponder, and pretending to write something on their page. But, for all the bluster and show, nothing was being achieved. Like the stallers, these students were hiding behind legitimate student behavior. The difference was that while the stallers hide behind legitimate off-task behavior, the fakers hide behind legitimate on-task behavior. When we interviewed them, we learned that, like the stallers, these students either didn't know how to do the task or were just killing time until the teacher went over it.

4. **Mimicking** - Unlike students in the three aforementioned groups, students who mimicked attempted, and often completed, the task. What they were doing, however, was trying to recreate the pattern of the solutions that had just been demonstrated on the board. This involved constant referencing to the demonstrated example with line-by-line mapping from the example to the task at hand. If the example that the teacher had demonstrated did not match the task they were asked to do, these students were often way off track or completely stuck. When we interviewed the teachers in whose classrooms we were doing the studenting research, all of them stated, with emphasis, that they did not want their students to mimic. Ironically, 100% of the students who mimicked stated that they thought that mimicking was what their teacher

wanted them to do. They were reading the demonstration of an analogous example prior to the now-you-try-one tasks as an invitation to mimic.

5. **Trying it on their own** - The last behavior was to just try it on their own. These students put their heads down and just tried to reason their way through the task based on their understanding. Some of them got it right, some of them got it wrong. Regardless, they were checking their understanding and getting feedback on it—as the teachers had intended.

These same five studenting behaviors were present every time we observed students in a now-you-try-one setting. And the distribution of how many students were exhibiting each behavior was surprisingly similar in each of the 10 classrooms in which we conducted this research (see Figure i.3). In all instances mimicking was exhibited by more than half of the class, with slacking, stalling, and faking combining to account for about a quarter of the students. Those trying it on their own—which is what the teacher wanted—only accounted for about 20% of the students. So, when I said that that I had a *sense* that students were not thinking, what I was actually seeing was slacking, stalling, faking, and mimicking—none of which is thinking.

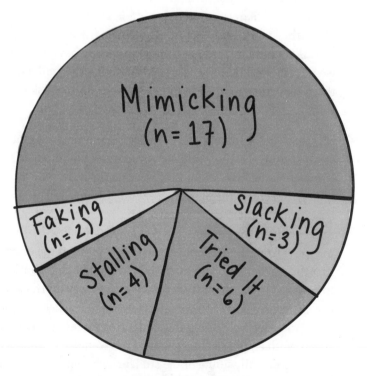

Figure i.3 Distribution of studenting behaviors on now-you-try-one tasks.

When I combined the studenting data for now-you-try-one tasks with the studenting data for note-taking (Chapter 11) and homework (Chapter 7), along with data from other activity settings, a clear picture emerged for exactly how much non-thinking behavior was present within a one-hour lesson. The results were troubling. In a typical one-hour lesson, 75%–85% of the students exhibited non-thinking behaviors for 100% of the time. The rest of the students exhibited non-thinking behaviors for all but 8–12 minutes of the time. This became my baseline data—the baseline from which I was hoping to make improvements.

INSTITUTIONAL NORMS

On my journey through these original 40 classrooms in 40 different buildings, other patterns began to emerge. Everywhere I went, irrespective of grade or demographic, classrooms looked more alike than they looked different. And what happened in those classrooms looked more alike than it looked different. There were differences, to be sure, but the majority of what I was seeing was the same. There were desks or tables, usually oriented toward a discernible front of the classroom. Toward this front was a teacher desk, some sort of vertical writing space for the teacher, and some sort of a vertical projection space. Students sat, while the teacher stood. Students wrote on horizontal surfaces while the teacher wrote on vertical ones. And the lessons mostly followed the same rhythm—beginning with some sort of teacher-led activity like a lecture or note-taking, perhaps shifting to some sort of small or big group discussion, but almost always culminating in some form of individual work. Even in the few more progressive classrooms I observed, the physical space looked the same, and the rhythm of the lesson was the same. What was different was the duration and nature of the activity in the middle of the lesson.

These normative structures that permeate classrooms in North America, and around the world, are so robust, so entrenched, that they transcend the idea of classroom norms (Cobb, Wood, & Yackel, 1991; Yackel & Cobb, 1996) and can only be described as institutional norms (Liu & Liljedahl, 2012)—norms that have extended beyond the classroom, even the school building, and have become ensconced in the very institution of *school*. Much of how classrooms look and much of what happens in them today is

> Much of how classrooms look and much of what happens in them today is guided by institutional norms—norms that have not changed since the inception of an industrial-age model of public education.

guided by these institutional norms—norms that have not changed since the inception of an industrial-age model of public education. Yes, desks look different now, and we have gone from blackboards to greenboards to whiteboards to smartboards, but students are still sitting, and teachers are still standing. And although there have been a lot of innovations in assessment, technology, and pedagogy, much of the foundational structure of *school* remains the same.

TOWARD A THINKING CLASSROOM

Everywhere I went I saw students not thinking and, as a result, teachers having to plan their teaching on the assumption that students either can't or won't think. And everywhere I went, I saw classrooms, and what happened in classrooms, that looked more alike than they looked different. So, I began to wonder if there were a connection between these in some way? Could the very institutional norms that permeate all schools and all classrooms actually be enabling and fostering the non-thinking behaviors I was observing? If this were true, what that would mean is that we would need to fundamentally alter the institutional norms to get students to think.

This assumption became the basis of my research, and for the next 15 years I worked with over 400 K–12 teachers to try to break through the non-thinking behaviors and get students to think. We worked in teams of 8–18 teachers in two-week cycles to deliberately break institutional normative structures and see whether it could increase student thinking. Our goal was simple—try to increase the number of students thinking and try to increase the number of minutes during which students were thinking. In essence, we wanted to improve on the baseline data. And we were willing to break any and all classroom norms to achieve it. Our only restrictions were that we would work within the confines of the classroom and within the confines of the set bell schedule. Other than that, there was no norm we were not willing to turn over.

To illustrate an extreme example of how far we were willing to go, early on in the research I worked with eight teachers who taught for two weeks in classrooms without any furniture. Furniture is an enduring institutional norm, and we wanted to see what would happen if we upended it. I learned three things from this experiment. First, student thinking increased—and radically so. We had more students thinking and thinking for longer. Despite this positive result, however, I also learned that teachers don't like to teach in classrooms without furniture. This realization was important and formed a structure for much of my research going forward. There is no point in researching a practice that teachers are unwilling to implement—irrespective of how positive the results are. This constrained the scope of what we were willing to try in the classrooms. This is not to say that we were not willing to push into spaces that were uncomfortable, but there were limits to what was reasonable.

The third thing I learned was that results often came before explanations. This remained true all through the research and continues to be true even today. Knowledge of what works always preceded an understanding of why it worked. As a researcher who is used to starting with theories and then testing them, this was new and exciting territory for me. In the case of no furniture, for example, it took many months of interviews with students in different contexts before I began to even get a glimpse of why having no furniture influenced student thinking. It turns out that when students walk into a classroom that looks like every other classroom they walk into, they assume that the lesson is going to go like every other lesson they have been part of. And, therefore, they bring all of their habits and studenting norms into the room with them. If those studenting norms are non-thinking behaviors, then they are going to not think in this lesson as well. When the students walk into a room that looks very different, however, then they leave their habits and norms at the door and allow themselves to be different—at least to begin with. The reason teaching in classrooms with no furniture had an effect on student thinking wasn't that it, in itself, promoted thinking but rather that it didn't trigger non-thinking habits. And this gave the teachers a chance to make something else happen. I will return to this idea in Chapter 15.

So, we launched into the research with enthusiasm, and almost immediately we started to see positive changes in student thinking. Teachers were reporting back great successes, and, when I would visit classrooms and gather data, I was seeing tremendous improvements in student thinking. In our enthusiasm to create change, however, we lost sight of what changes were having what impact. We were trying so many things at once that we lost control of cause and effect— pedagogy and thinking. We needed to be more systematic in our experimentation. We needed to pick one variable to experiment with for two weeks and measure the effects on student thinking through that one variable. But what were the variables?

The obvious choice was the list of activity settings I had studied during the studenting research—now-you-try-one tasks, notes, homework, review, group work, et cetera. But the list of what influences thinking in a classroom goes well beyond the discrete moments in a lesson. For example, I have already demonstrated that how a room looks when students walk in has an impact. So too do how we ask and answer questions, the types of tasks we use, and so on.

In an effort to find a list of variables that impact thinking in a classroom, I spent several months visiting classrooms that I was not, at the time, running experiments in. I was looking for a way to disaggregate teaching into discrete factors, each of which could act as a variable in our pursuit to improve thinking in the classroom. In the end, a list of 14 such factors emerged.

1. What types of tasks we use
2. How we form collaborative groups
3. Where students work
4. How we arrange the furniture
5. How we answer questions
6. When, where, and how tasks are given
7. What Homework looks like
8. How we foster student autonomy
9. How we use hints & extensions
10. How we consolidate a lesson
11. How students take notes
12. How we choose to evaluate
13. How we use formative assessment
14. How we grade

This list is comprehensive. Everything we, as teachers, do in the classroom is an enactment of one of these factors, and how we enact each of these factors is what forms our teaching practice—our unique teaching practice.

These factors became the variables we systematically experimented with in our efforts to increase thinking in the classroom. What we were looking for were practices, for each factor, that generated more thinking than the institutionally normative practices I had observed. And of these practices, we were looking for the practices that generated the most thinking—what we eventually came to call the *optimal practice for thinking*. And we found them. Slowly at first. But over the next 15 years they all emerged.

As it turned out, finding practices that generated more thinking than the institutional normative practices was not difficult. The normative practices were far from optimal, and there are many ways to enact each of the 14 factors such that they generate more thinking. In most cases we began our research by enacting a practice that was the exact opposite to what the norm was—if the norm was that students sit, then we made them stand; if the norm was that we answer students' questions, then we stopped answering questions; and so on. In some cases, this contrarian approach produced the optimal practice, but in all cases, it produced a practice that generated more thinking than the baseline data.

Groups of teachers tried each practice for two weeks. If it produced good results, then we tweaked it, and the teachers kept going with it. If, along the way, we tried a practice that was less effective than another practice we had tried, we abandoned it and tried something else. And so on. Eventually, after a number of iterations, we would get to the point where any changes we made to the practice made it less effective. At that point we had what I called a *local optimal practice*—it was optimal for that particular teacher, in their particular setting, with their particular demographic of students. Although these practices were of interest for teaching in general, they were often intertwined with aspects of the teacher's personality, habits, and norms. What I really wanted were practices that worked for any teacher in any setting.

So, I would take these local optimal practices and give them to different teachers in completely different settings, teaching different demographics of students, and see how these practices worked for them. Then we would run two-week cycles of iterations among those teachers, until what emerged was a practice that produced the most thinking and was transferable across teachers, settings, and demographics. I would then give that practice to a new group of teachers to use for six to eight weeks to see if it had longitudinal fortitude and was not just something that worked because it was new to students. If it passed this last hurdle, then this practice was now what I was willing to consider an *optimal practice for thinking* within the factor we had experimented with.

HOW TO READ THIS BOOK

In the chapters that follow, you'll read about each of the 14 optimal practices for thinking that emerged from the research into each of the 14 aforementioned variables. Each chapter begins with a brief description of which factor the chapter is addressing, why it is important, and what you will learn in that chapter. This is followed by an exploration around **The Issue** concerning the institutionally normative practices for this factor and what is **The Problem** that comes with these normative practices.

These introductory sections are then followed by the main part of the chapter, called **Toward a Thinking Classroom**, where you'll learn about the optimal thinking practice for the factor in question and how this practice generally addresses some of the problems raised in the introductory sections, along with some grade-band or demographic-specific guidance where there is nuance. This is also the section in which you'll encounter a lot of concrete advice for implementing these practices. In our research into the optimal practices for thinking for each factor, what emerged were a number of what I came to call *micro-moves*. These are the little things within each of the practices that we found enhanced, streamlined, or made easier to implement the optimal practice. These are called micro-moves to contrast them against the *macro-moves* that are the optimal practices for thinking in each chapter. This is not to say they are any less important. In many cases, these micro-moves make the difference between smooth and rough implementation in your classroom.

Some of the things you read in **The Issues** and **The Problem** sections of each chapter will likely disturb you, as you may read about problems with practices that you are using. You may feel challenged by those ideas, and you may have questions about them. At the same time, some of the results you read in the **Toward a Thinking Classroom** sections may be difficult to imagine, and you may have questions about them or how to implement them in your classroom. As such, the next section in each chapter is called **FAQ**—frequently asked questions. This section addresses the questions that I find educators are most often curious about. I hope that the questions I address are the same questions that arise for you as you read the chapter.

Each chapter ends with a quick summary of the **Macro- and Micro-Moves** and a series of **Questions to Think About**. These questions can be used as discussion points if you are reading this book as part of a professional learning community (PLC), if you are in a methods course, or in partnership with another teacher. If you are reading the book by yourself, these questions can also be used to push you to think more deeply about what you have read in the chapter and how what you read will translate into your classroom. Some of the questions are also designed to help you uncover some of the implicit beliefs that you have about teaching mathematics that could be the source of some of your challenges with or disbelief of what is presented in the chapter.

The book is written in such a way that you can read the whole book before you begin to build your own thinking classroom. If this is how you choose to engage with the book, then Chapter 15 will provide the results of the research into the optimal sequence for implementation and which practices need to be implemented together. If you want to build your thinking classroom as you read each chapter, then the book is also written to accommodate that. If this is how you choose to engage with the content, I suggest that you read Chapters 1–3 and then implement all three of those optimal practices for thinking together. After that, you can implement each practice as you read about it. To help you along the way, each chapter ends with a **Try This** section where you are provided with some tips and tricks as well as thinking tasks that you can use to help initiate that thinking practice in your classroom.

This is not to say that you must implement each optimal practice exactly as stipulated in the chapter. These practices are a framework that is meant to come alongside your current teaching experience. All of your teacherly craft is still relevant and necessary to make each of these optimal practices work in your classroom. The micro-moves will help. And as you enact each practice within your particular setting and with your particular demographic, you will find new micro-moves that allow you to make each practice even better.

Enjoy the journey.

CHAPTER 1

WHAT TYPES OF TASKS WE USE
IN A THINKING CLASSROOM

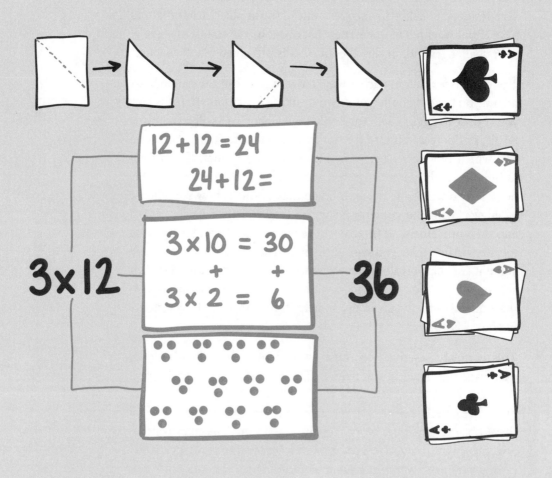

If we want our students to think, we need to give them something to think about—something that will not only require thinking but will also encourage thinking. In mathematics, this comes in the form of a task, and having the right task is important. So, while the rest of the book will look at the things we can do in our teaching practice to build thinking classrooms, this chapter will look specifically at the tasks around which thinking classrooms are built. By the end of this chapter you will have learned about the different types of tasks that you can use to build a thinking classroom, where to find them, and how to design your own.

> If we want our students to think, we need to give them something to think about.

 ## THE ISSUE

Tasks are inert. To come alive, they need an audience to solve them. So, when I talk with teachers about what makes a good task for building thinking classrooms, I don't talk about what a task is, but rather what a task does. And what a task needs to do is to get students to think. Consider, for example, the following task:

Which is greater, eight or nine?

You may be thinking that this is not a good task. And if this question were posed to Grade 9 students, you would be correct. That is the wrong audience for this task. But if this same question were asked of a four-year-old child, this turns out to be a very good task. The strategies that the child would need to invoke in order to figure this out are both complex and nuanced and would require a lot of thinking to resolve. So, the question is not whether *which is greater* is a good task or not. The question is, what is it good for? And the answer to that question is that it is good for getting students, for whom the relative cardinality and/or positionality of the number symbols have not yet been routinized, to think.

When it comes to talking about tasks that get students to think, the best place to start is with problem solving. From Pólya's (1945) *How to Solve It* to the *NCTM Principles and Standards* (2000), the literature is replete with the benefits of having mathematics students engage in problem solving. Although there are arguments about the exact processes involved and the exact competencies required, there is universal agreement that problem solving is what we do when we don't know what to do. That is, problem solving is not the precise application of a known procedure. It is not the implementation of a taught algorithm. And it

> Problem solving is what we do when we don't know what to do.

is not the smooth execution of a formula. Problem solving is a messy, non-linear, and idiosyncratic process. Students will get stuck. They will think. And they will get unstuck. And when they do, they will learn—they will learn about mathematics, they will learn about themselves, and they will learn how to think.

> Good problem-solving tasks require students to get stuck and then to think, to experiment, to try and to fail, and to apply their knowledge in novel ways in order to get unstuck.

As with good tasks for building thinking classrooms, what makes a good problem-solving task is based on what it does—or rather, what it requires students to do to solve it. Good problem-solving tasks require students to get stuck and then to think, to experiment, to try and to fail, and to apply their knowledge in novel ways in order to get unstuck. The *cats and rats* problem in the introduction is a good example of such a task. Knowledge of fractions and ratios is necessary, but far from sufficient, to solve this problem. Yet, no other mathematical content knowledge is needed. To solve it—to get unstuck—we need to think about the problem differently than we usually think about equivalent fractions or common ratios. We need to come to the realization that if six cats kill six rats in six minutes, then either six cats will kill one rat in one minute, or one cat will kill one rat in six minutes. How a student gets to this realization is problem solving.

Problem-solving tasks are often called non-routine tasks because they require students to invoke their knowledge in ways that have not been routinized. Once routinization happens, students are mimicking rather than thinking—or as Lithner (2008) calls it, being imitative rather than creative. Good problem-solving tasks are also rich tasks in that they require students to draw on a rich diversity of mathematical knowledge and to put this knowledge together in different ways in order to solve the problem. They are also called rich because solving these problems leads to engagement with a rich and diverse cross section of mathematics. Regardless of how they are referred to, what makes a task a good problem-solving task is not what it is, but what it does. And what they do is make students think.

My early research into building a thinking classroom was very much focused on tasks. Despite my experiences in Jane's class, I still believed that the best way to get students to think was to give them a task that would motivate, even necessitate, them to think. For this reason, I spent a lot of time searching for and designing tasks that would do just that. What emerged from these efforts was a collection of what I started out calling *highly engaging thinking* tasks. To this collection I also added a lot of mathematical *card tricks* and developed a genre of real-world problem-solving tasks that I called *numeracy tasks*.

Let's take a closer look at each of these three kinds of tasks:

1. **HIGHLY ENGAGING THINKING TASKS** are so engaging, so interesting, that people cannot resist thinking. They have broad appeal and can be used across a wide range of grades, with some being able to be used all the way from Grade 4 up to calculus and beyond. At first, I thought they were rare—so rare, that for a long time I didn't know if they really existed. And then I found one. And then another. And then several. Now I realize they are plentiful if you know where to look. Here are four examples of such tasks, organized by grade band:

   - **PRIMARY:** How many squares are in the image below?

   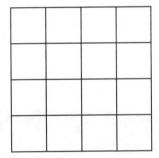

   - **INTERMEDIATE:** I buy a video game for $10. I then sell it for $20. I buy it back for $30. Finally, I sell it again for $40. How much money did I make or lose?

   - **MIDDLE SCHOOL:** I have a four-minute egg timer and a seven-minute egg timer—the kind that you turn over and let the sand run through. Can I use these to cook a nine-minute egg? If so, how long will someone have to wait for their egg?

   - **HIGH SCHOOL:** An eccentric woman has booked three adjacent and adjoining hotel rooms. When she checks in, she tells the receptionist that if he needs her, she will always be in the room next door to the room she was in the night before. The receptionist thinks nothing of this until an hour later when he realizes that her credit card has been declined, and he must now go find her. The problem is that he is very busy and only has time to knock on one door per day. How many days does he need to guarantee that he finds her? What if it were four rooms? Five rooms? What if it were 17 rooms, and she is checked in for 30 days—can he find her before she leaves?

I will share more of these types of highly engaging thinking tasks throughout the book, beginning with some at the end of Chapter 3.

2. **CARD TRICKS** have the same qualities as highly engaging thinking tasks—they are highly engaging situated tasks that draw students in and entice them to think. It turns out that there are a lot of card tricks that are both built on and can be explained by mathematics. These were the ones I was interested in. What I was not interested in were card tricks that relied on sleight of hand. I wanted students to engage with the magic of mathematics, not the magic of my hands. Video 1.1 shows an example of one of these tasks. If you are interested in these kinds of card tricks, you can find a collection of them on my website (http://www.peterliljedahl. com/teachers/card-tricks).

VIDEO 1.1

If you are interested in these kinds of card tricks you can find a collection of them on my website (http://www. peterliljedahl.com/ teachers/card-tricks).

Source: Youtube video via peterliljedahl.com

3. **NUMERACY TASKS** are tasks that are based not only on reality, but on the reality that is relative to students' lives. From cell phones to entertainment to sports, these tasks are built up specifically to engage students in rich tasks wherein they have to negotiate the ambiguity inherent in real-life experiences. For example,

SKI TRIP FUNDRAISER

The ski club is finally going skiing. Each person tried their best to raise money for their trip. Below is a chart that shows how much money each person raised, and their individual cost, depending on whether they need rentals or lessons. All of the money raised must be applied to the cost of the trip, and every person must go on the trip, even if it means that they may have to put in their own money to do it. Have they raised enough? If not, who needs to pay, and how much do they need to pay?

Name	Amount Raised	Rental Cost	Lift Ticket	Lesson Cost
Alex	75	20	40	40
Hilary	125	10	40	40
Danica	50	30	40	0
Kevin	10	40	40	40
Jane	25	0	40	0
Ramona	10	0	40	40
Terry	38	30	40	0
Steve	22	40	40	40
Sonia	200	20	40	0
Kate	60	25	40	0

More examples of these tasks, and how they are made, can be also be found on my website (http://www.peterliljedahl.com/teachers/numeracy-tasks).

Low-Floor Task: Task with a threshold that allows any and all learners to find a point of entry, or access, and then engage within their level of comfort.

High-Ceiling Task: Tasks that have ambiguity and/or room for extensions such that students can engage with the evolving complexity of the task.

Open-Middle: A problem structure where a task has a single final correct answer, but in which there are multiple possible correct ways to approach and solve the problem.

All three of these types of tasks provide engaging contexts that draw students in and entice them to think. Therefore, these tasks are useful in building thinking classrooms. Aside from context, all these tasks also have easy entry points (low floor) and evolving complexity (high ceiling), and they drive students to want to talk and to collaborate.

Whereas the inherent ambiguity of numeracy tasks makes them truly open ended—with some having as many as 200 viable and defendable solutions—the highly engaging thinking tasks and card tricks usually have only one final answer. However, they allow for multiple approaches to get to that one answer and, hence, have an open-middle structure.

THE PROBLEM

Aside from being rich and engaging tasks with the ability to get students to think, these aforementioned tasks share another quality—they are, for the most part, all non-curricular tasks. That is, very few of these tasks require mathematics that map nicely onto a list of outcomes or standards in a specific school curriculum. Consider, for example, the difference between two tasks that can be used with Grade 8 students: the *True or False* card trick in Video 1.1 and a task that asks students to add two proper fractions with different denominators. The *True or False* task is clearly mathematical in nature; the solution to it requires that students attend to the position of the target card, the patterns in the cardinality of the number of letters in certain words, and the role that reversing order plays—none of which is an outcome in a Grade 8 curriculum. On the other hand, asking students to add two fractions with different denominators requires them to understand that a common denominator is needed, be able to find the lowest common denominator, add fractions, and potentially be able to reduce a fraction—all of which are outcomes in some Grade 8 curricula. So, whereas both tasks are mathematical in nature, the *True or False* card trick is non-curricular, while the adding-fractions question is curricular.

Non-curricular Task: A task that is clearly mathematical in nature but does not map well to the outcomes or standards specified in the curriculum for the class in which it is used.

Even if there is a rich task that maps nicely to the curriculum you are teaching, it only maps to curricular outcomes if students happen to solve the problem using concepts and skills from their current curriculum. This is the nature of open-middle and open-ended tasks. Such tasks invite students to think for themselves. And when students begin to think for themselves, a lot of unpredictable things can happen. If your goal is only to get students to think, then this is not a problem. If your goal is to use a rich task to, for example, get students to think about division of fractions, then this can be a problem. Of 30 students, only a handful may choose a solution path that follows the lines of curriculum you were hoping a rich task would touch on. The rest may choose to use repeated subtraction, repeated addition, or a type of logic that makes unnecessary the need to think about fractions at all. Depending on the grade you are teaching, these solution paths, although not achieving what you were hoping for, may still touch on topics from your curriculum. More often, however, this is not the case.

If, in reaction to this, we try to force a more predictable curriculum mapping by artificially constraining tasks, before long we have reduced what was once a rich task to the type of word problem we often see in mathematics textbooks:

> Camille went to the store to buy eggs, milk, and cheese. Eggs cost $3.50, milk costs $2.00, and cheese costs $4.00. How much money did Camille need?

Word problems, like rich tasks, require the student to decode what is being asked. However, once a word problem is decoded, the mathematics is often trivial, procedural, and analogous to the mathematics that was taught that day. This is not true of rich problem-solving tasks. In a rich task, once the language has been decoded, the mathematics that is needed to solve it is neither trivial nor procedural. Basically, in rich tasks the problem is in the mathematics, and in word problems the problem is in the words—this is maybe why they are called word problems.

> Once a word problem is decoded, the mathematics is often trivial, procedural, and analogous to the mathematics that was taught that day. This is not true of rich problem-solving tasks.

Whereas rich tasks get students to think at the expense of meeting curriculum goals, word problems more predictably and reliably push students to use specific bits of learned knowledge—but often at the expense of engagement and the thinking that we need to foster in our students. So, how then do we move forward from this reality?

TOWARD A THINKING CLASSROOM

One way forward, although seemingly unrealistic, is to stop worrying about curriculum. My earliest efforts to build thinking classrooms did just this. Rather than think about curriculum, I was only concerned with getting students to think. This is not to say that I was naïve about the lived reality of classroom teachers and the persistent and ubiquitous nature of curriculum. Rather, it is just that I needed to start somewhere. Before I could even begin to think about how to get students to think about curriculum, I needed to get students to think.

This proved to be surprisingly easy. Once we shed the burden of curriculum, it turns out that there are a huge number of resources available to us that are effective for getting students to think. From problems of the day to brainteasers, the internet is full of resources that are engaging and thought provoking. Some of these, it can be argued, address curriculum—but, again, only for those students who follow a particular solution path.

Students, as it turns out, want to think—and think deeply. My early efforts to build thinking classrooms through the use of highly engaging thinking tasks, card tricks, and numeracy tasks—and my cavalier attitude about curriculum—were actually hugely successful. Successful to the point where I could give a teacher a set of three tasks and, without any other changes, could dramatically increase both the number of students who were thinking and the number of minutes that were spent thinking. On top of that, students were enjoying and looking forward to mathematics and the next task, their self-confidence and self-efficacy increased, and they became better mathematical thinkers.

> Students, as it turns out, want to think—and think deeply.

Figure 1.1 Students in an elementary classroom engage in a thinking task.
Source: FatCamera/iStock.com

The trick was to maintain the positive effect, and positive affect, while turning our attention back to the reality of curriculum. To do this, I had one thread to follow—the thread that comes from the understanding that problem solving is what we do when we don't know what to do. Curriculum tasks are typically the exact opposite of this. Curriculum tasks are often what students do when they know what to do—*after* they have been shown how. Asking a high school student to factor $x^2 - 5x - 14$ or an elementary student to solve $3.1 + 5.2$ after they have been shown how promotes mimicking, not thinking. My observation of those initial 40 classrooms showed that this is exactly when and how curriculum tasks were most often used.

Having said that, it turns out that both of these questions are excellent thinking questions—if they are asked *before* the students have been shown how to answer them. Herein lay the root of how to get students to think while at the same time addressing grade-specific curriculum. For example, let's look more closely at the factoring quadratic task and how that question can be presented without first teaching students how to do it.

> Asking a high school student to factor $x^2 - 5x - 14$ or an elementary student to solve $3.1 + 5.2$ *after* they have been shown how promotes mimicking, not thinking.

Teacher   Let's start with a bit of review. How would I expand $(x + 2)(x + 3)$?

[Teacher writes on the board $(x + 2)(x + 3) =$]

Students $x^2 + 5x + 6$.

[Teacher writes on the board $(x + 2)(x + 3) = x^2 + 5x + 6$]

Teacher   OK. So what if my answer were $x^2 + 7x + 6$? What would the question be?

[Teacher writes on the board ( )( ) = $x^2 + 7x + 6$ right underneath the previous line.]

For adding decimals, the question could be posed in a similar fashion.

Teacher   Let's start with a brief review. Can someone tell the class what 3.1 means?

Student   This is a number that is bigger than 3 but less than 4.

Teacher   Is it closer to 3 or 4?

Student   It is closer to 3.

Teacher   OK. And what is 5.2?

Student   It is a number between 5 and 6 that is closer to 5.

Teacher   OK. If I add 3.1 and 5.2, what two whole numbers is the answer between, and which number is it closer to? What would the answer be?

Even counting at the primary level can be turned into a thinking task.

Teacher   Let's all count together up to 20.

Students 1, 2, 3, 4, 5, . . ., 20.

Teacher   Ok. What if we start at 14? What are the three numbers that come after 14? What are the three numbers that come before 14?

These scripts are similar in that they begin by asking a question about prior knowledge, then they ask a question that is an extension of that prior knowledge, and they ask students to do something without telling them how. And, as such, they require students to think, not only in general, but also about particular curriculum. It turns out that almost any curriculum tasks can be turned from a mimicking task to a thinking task by following this same formulation—begin by asking a question that is review of prior knowledge; then ask a question that is an extension of that prior knowledge.

In my research, I compared three types of lessons (Figure 1.2).

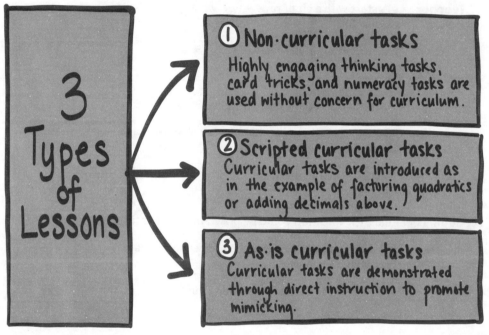

Figure 1.2 Three types of lessons.

There were big differences between how students performed in these types of lessons. Although the first two lesson types were both designed around tasks to get students to think, the lesson that was designed around non-curricular tasks (Type 1) got many more students to think than the lesson scripted to get students to think about curriculum (Type 2). Simply turning a standard curricular task into a thinking task was not enough to get all the students thinking.

Similarly, whereas the second and third types of lessons are both built around curriculum tasks, the lesson where direct instruction was used (Type 3) allowed more students to successfully complete the task at hand. This is not surprising, as mimicking can be an effective strategy that may allow students to be successful in the short term.

But, as mentioned in the introduction, mimicking is not thinking and therefore not learning. Naturally, there were also students who were successful on the rescripted curricular thinking task (Type 2). There were just fewer than in the direct instruction lesson.

However, an interesting thing happened when three lessons using non-curricular tasks (Type 1) *preceded* students' exposure to the scripted curricular thinking task (Type 2)—the number of students who successfully completed the scripted tasks (Type 2) surpassed the number who were successful in the mimicking lessons (Type 3). In other words, students can be successful at these types of scripted thinking tasks, even more successful than in lessons designed to promote mimicking, if their willingness to think is *first* primed with the use of good non-curricular tasks. This makes sense. Type 1 tasks are more likely to engage students with their rich and interesting contexts and propel them into thinking than a task asking them to think about factoring quadratics, adding decimals, or counting. But once the thinking starts, it becomes an end unto itself, and students are not only more willing to think but they want to think. The non-curricular tasks (Type 1), in this regard, served as a primer for—and thus made room for—the more curriculum-driven scripted thinking tasks (Type 2).

> Once the thinking starts, it becomes an end unto itself, and students are not only more willing to think but they want to think.

Further investigation showed that although three lessons of non-curricular tasks (Type 1) was enough to prime many classes, in some cases as many as five lessons were needed before the dispositions of the students shifted enough to allow them to be successful at scripted curricular thinking tasks. This investigation revealed that, in almost every situation, the teacher was able to predict when the class was ready to shift their thinking toward curricular thinking tasks.

Lucy	I don't know why, but they just seemed ready. There was no more whining, and the kids came into class excited about seeing the problem they would work on that day.

The key was, however, that in the transition from a non-curricular task (Type 1) to a curriculum thinking task (Type 2), nothing else changed. The teacher posed the task as a challenge—as a problem to solve—without any big declarations that now we are going to start doing curricular tasks in a different way.

This is not to say that all students were successful or that all students were willing to think. Far from it. Simply turning a basic curriculum task into a thinking task does not mean students are automatically going to think. More things need to change in the lesson if thinking is to be built and sustained over time, and that is what the rest of this book is about. However, these results show that to get students thinking *about* curriculum tasks, they need to first be primed to do so using non-curricular tasks. Nothing in my research has shown a way to avoid this. You have to go slow to go fast.

In Chapter 9, I will discuss much more about how to build a sequence of scripted curricular thinking tasks (Type 2) that follow on the heels of the aforementioned engaging non-curricular tasks (Type 1) and allow students to effectively think their ways through large amounts of curriculum quickly. For now, however, it is sufficient to say that the goal of this book is not to get students to think about engaging non-curricular tasks day in and day out—that turns out to be rather easy. Rather, the goal is to get more of your students thinking, and thinking for longer periods of time, within the context of curriculum.

Q In this chapter, as well as the introduction, mimicking is portrayed as something bad. Isn't mimicking a good starting point for students before moving onto thinking tasks?

A The question is not whether mimicking is good or bad. The question is, what is mimicking good or bad for? Mimicking is very good at teaching students how to replicate routines—the routine for factoring quadratics, adding decimals, dividing fractions, et cetera. So good, in fact, that once students start to have success with mimicking, they don't want to stop. Mimicking is an addiction that is easily acquired at lower grades and difficult to give up at higher grades. You may have seen this when trying to explain a difficult concept and some of your students are asking you to "just show us how to do it." The problem is that mimicking is only an effective strategy when the number of routines to memorize is small. As the student moves up in grades, the number of routines per topic increases, until this becomes an unmanageable and ineffective strategy. Yet students who have had success with it in the past are resistant to abandoning it. Furthermore, mimicking tends to create short-term success without the long-term learning that allows students to make connections with other topics in the same and subsequent grades. So they do not develop the web of connections that helps them understand mathematics.

Mimicking is bad because it displaces thinking. Mimicking happens not alongside, but instead of, thinking. Likewise, mimicking is not a precursor to thinking. Mimicking requires less energy and less effort than thinking, and once the mimicking has begun, it is difficult to ask students to shift their attention to something that takes more time, more energy, and more effort. Our research on studenting and homework showed that only 20% of students who mimicked at the beginning of their homework assignment were even willing to attempt questions for which they did not have an analogous worked example and that would require them to think. And of those, only half were able to complete a question for which they did not have an analogous example in their notes.

Q I don't have time to give up three to five days of my school year to do non-curricular tasks. Can't I just jump right in with curriculum thinking tasks?

A Starting to build thinking classrooms with non-curricular tasks is imperative. As already mentioned, their use dramatically increases your students' success with scripted curriculum thinking tasks when you transition to those types of tasks. How it does this has not been mentioned, however. Well selected non-curriculum tasks, with their engaging contexts, propel students to want to begin to think. They create situations where every student gets stuck, which makes stuck an expected, safe, and socially acceptable state to be in. In essence, these tasks make it safe to fail and keep trying. And through these struggles, students begin to build confidence in their teacher's confidence in them. All of these qualities are easier to build inside of highly engaging non-curricular tasks and are necessary when we transition students to curricular thinking tasks.

This is not to say that these same qualities can't be built inside of curricular thinking tasks, but it is harder, takes longer, and will only work well with a few students. Curricular tasks are too familiar to signal that something has changed, and thereby are less likely to prompt a change in behavior.

Q If non–curricular tasks—especially highly engaging thinking tasks—are so good at engaging students, why don't we just teach all of mathematics that way? There must be a collection of tasks, the whole of which will cover an entire curriculum.

A This is a bold approach, which has been proven to work. This is the essence of Jo Boaler's early research at Phoenix Park (Boaler

2002). Further, Maria Kerkhoff (2018) showed that after doing just 18 rich tasks over the course of 18 classes, the student who she was studying encountered almost all of the curriculum outcomes for her grade, along with numerous curriculum outcomes from previous and future grades. In essence, if we just get students thinking about lots of different problems, the curriculum outcomes will eventually be covered, irrespective of which solution paths students follow. This is the approach a group of mathematics educators in Alberta took. They have created collections of tasks for Grades 2, 3, and 8, which allow them to cover all of the curriculum. You can access these collections by going to Alicia Burdess's website (http://www.aliciaburdess.com/teaching-through-problem-solving.html).

The problem is that such a move takes a lot of faith on the part of the teacher. And this faith is quickly eroded if there are set dates by which students must have learned certain concepts. The other issue is that the higher the students get in the grades, the more difficult it becomes to find collections of non-curricular highly engaging thinking tasks that will, in their entirety, cover curriculum—the more abstract mathematics gets, the more difficult it becomes (not impossible) to create such resources.

Q Even if I want to use curricular thinking tasks, it will take so much longer to have students think their way to solutions than if I just show them. How will I find the time for that?

A There are a lot of aspects of time that came out in the research. First and foremost is the time it takes before students are given an opportunity to answer a question on their own. In lessons designed around having students mimic (Type 3), this opportunity does not occur until 15–35 minutes into the lesson. When using thinking curricular tasks, this happens in a fraction of that time. Looking back at the three sample scripts in this chapter, you will notice they are all brief. Very brief. I will discuss more in Chapter 6 how important this is. For now, however, it is enough to say that when relying on previous knowledge to prompt thinking, these types of scripts will always be brief.

The second aspect of time is how long it takes students to solve a task when asked to think versus when they are asked to mimic. In each of the example scripts, not only is the set-up quicker, the students tend to come to an answer more quickly. This may not be true the first time you design a curricular thinking script, but it goes faster and faster the more adept the students become at thinking.

Finally, my research shows that when curricular thinking tasks are combined with the other 13 practices, students move through a lot of content very quickly. The script for factoring quadratics, for example, when used in a fully implemented thinking classroom context, will cover the entire unit on factoring quadratics in 40–70 minutes. Adding and subtracting decimals takes less. I will discuss this more in Chapter 9. For now, however, it is sufficient to say that yes, it will take more time in the beginning, but you will earn all that back as your classroom becomes a thinking classroom.

Q Can students really solve curricular thinking tasks (Type 2) without first being shown how to do them?

A Yes. Even when these tasks were introduced on their own, students who were willing to think were generally successful at solving them. But not everyone was willing to think. Using highly engaging non-curricular tasks as a precursor to the curriculum thinking tasks increased dramatically the number of students who were willing to think while at the same time increasing the amount of time that all students were willing to think for—both of which will lead to more students being successful at solving the tasks.

Q For a curricular task to generate thinking, it should be asked before students have been shown how to solve it. Does this mean the task should come right at the beginning of the lesson?

A Yes. In Chapter 6 I will more thoroughly discuss how important this turned out to be. In the meantime, suffice it to say that thinking tasks should be asked in the first five minutes from the time you begin the lesson.

Q Each of the examples in this chapter drew on prior knowledge. What does it look like when we are starting with an entirely new topic, a topic for which the students have no prior knowledge?

A Curriculum is inherently spiraled. For this reason, it is seldom the case that students have no prior knowledge at all. In the rare cases where it is true, however, you can, if you wish, just tell the students something. But you still only have five minutes before you should ask them a thinking question. Take for example, the introduction of the Pythagorean Theorem. I offer two different scripts that can be used, the first of which relies on pattern spotting.

Teacher I am handing out a sheet with eight different triangles [see Figure 1.3], each with all its side lengths indicated. What sorts of patterns do you notice?

Figure 1.3 Pythagoras sheet.

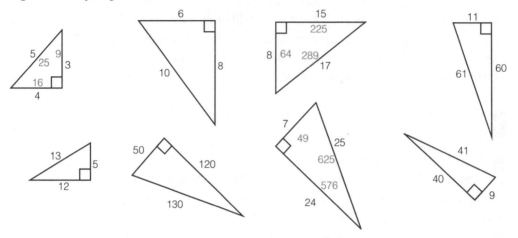

A sheet structured as in Figure 1.3 would allow students to notice that all the triangles are right triangles. They may also notice that some of the triangles are proportional to each other. They may notice that the extra numbers on some of triangles are the squares of the sides. Finally, they may notice that there is a relationship among these square numbers.

The second script involves a more direct approach.

Teacher If you look at the three triangles I have drawn here, you will notice that they are all right triangles. All right triangles share the property that the sum of the squares of the shorter two sides equals the square of the longer side. This is called the Pythagorean Theorem, and it is written as $a^2 + b^2 = c^2$, where a and b are the lengths of the shorter sides, and c is the length of the longer side. For example, we see that in the first triangle $3^2 + 4^2 = 5^2$. In the second triangle we see that $5^2 + 12^2 = 13^2$. Knowing this, consider this third triangle, where the shorter two sides are 8 and 15. What must the longer side length be?

Although very different in approach, both of these scripts have students doing a question they have not been shown how to do in the first five minutes. The first script promotes pattern spotting, while the second approach asks them to apply a known property. Regardless, they are going to have to think their way forward.

Q So, I run the script, and the students successfully answer the thinking question I pose. What do I do next?

A You ask a similar but more difficult question. All of Chapter 9 is about this, but for now just ask progressively harder questions. For example, in the second script above, you may ask the students to answer a question where the two shorter sides are 3.4 and 5.2 units long. Then ask them to answer a question where they are given the lengths of the longest and one of the shorter sides, et cetera.

SUMMARY

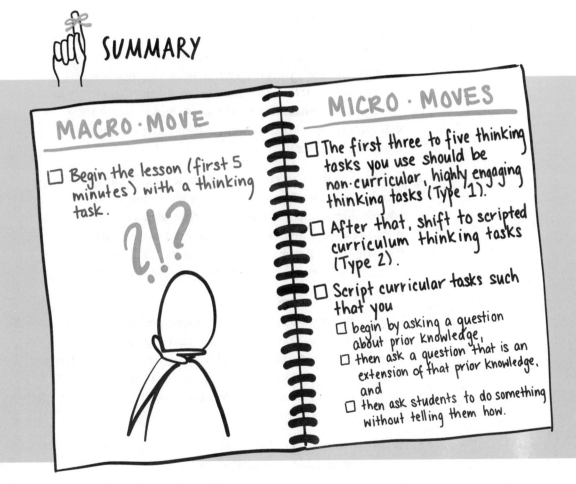

MACRO · MOVE

☐ Begin the lesson (first 5 minutes) with a thinking task.

MICRO · MOVES

☐ The first three to five thinking tasks you use should be non·curricular, highly engaging thinking tasks (Type 1).

☐ After that, shift to scripted curriculum thinking tasks (Type 2).

☐ Script curricular tasks such that you
  ☐ begin by asking a question about prior knowledge,
  ☐ then ask a question that is an extension of that prior knowledge, and
  ☐ then ask students to do something without telling them how.

QUESTIONS TO THINK ABOUT

1. What are some of the things in this chapter that immediately feel correct?

2. In this chapter you read about the negative consequences of mimicking. Can you think of any positive benefits? If so, do these positive benefits outweigh the negative consequences?

3. The introduction mentioned that almost all students who mimic express that they thought this is what they were meant to be doing. This chapter shares that one of the ways in which students come to this conclusion is by having their teachers show them how to do something before asking them to try it on their own. What other ways may we be communicating that mimicking is what we want students to do—even if that is not what we want?

4. You have read in this chapter that curriculum is inherently spiraled and, therefore, there are very few examples where you would introduce a topic for which students have no prior knowledge upon which such a script can be built. Can you think of some examples of such situations in your curriculum? If you can, is there really no prior knowledge that can be drawn on?

5. In this chapter it was shown that students perform better on scripted curricular tasks if they have first experienced three to five classes of working on highly engaging non-curricular tasks. How do you feel about giving up this time? What are the barriers for you to do this? What do you stand to gain? What do you stand to lose?

6. What are some of the challenges you anticipate you will experience in implementing the strategies suggested in this chapter? What are some of the ways to overcome these?

☑ TRY THIS

As mentioned in the introduction, the ideas in the first three chapters are best implemented together. Of course, you can ignore this and implement the ideas in this chapter right now. If you are doing this, remember to start with three to five non-curricular tasks and to get students doing these in the first five minutes. If, however, you are going to heed the advice and wait until the end of Chapter 3 to try anything with your students, then this is the time to create some scripts in preparation for this.

This chapter included three examples (counting, adding decimals, and factoring quadratics) of how to script the introduction of a task so that you can ask students to think their way through a problem without first showing them how to do it. These examples are all predicated on the idea of asking the students a question about prior knowledge, and then asking a question that is an extension of that prior knowledge. Consider some topics you have recently taught or are about to teach, and create some scripts for these topics.

CHAPTER 2

HOW WE FORM
COLLABORATIVE GROUPS
IN A THINKING CLASSROOM

We know from research that student collaboration is an important aspect of classroom practice, because when it functions as intended, it has a powerful impact on learning (Edwards & Jones, 2003; Hattie, 2009; Slavin, 1996). You are likely already using groups in your classroom to some degree, but are you satisfied with the engagement you see in your students? Are they *all* participating in the ways you would hope? What kind of grouping practices work *best* for collaboration, or could make your own collaborative grouping efforts work *better*? Our research shows that the answer may not be what you think. In this chapter we are looking closely at how the way in which we group students for collaborative problem solving actually impacts the way students *engage* in the collaborative effort. By the end of the chapter you will have a method and rationale for grouping students that will drastically shift student engagement, participation, and community within your room.

 ## THE ISSUE

In all the classrooms I've researched and observed, one thing that always struck me was the number of teachers who incorporated collaboration into their teaching in one way or another. This generally ranged from simply having students sit in pairs and conducting turn-and-talks to having students work in assigned or self-selected groups for parts of the lesson. In elementary classrooms, I most often observed teachers using the *strategic grouping* method, where the teacher carefully arranges homogenous or heterogeneous groups to meet either the *educational* or the *social* goals for the class (Dweck & Leggett, 1988; Hatano, 1988; Jansen, 2006).

In our observations, we found that grouping for educational goals generally follows one of three rationales:

1. **PEDAGOGY:** Because a teacher believes that their students can, and will, learn from each other, the teacher will create either homogenous or heterogeneous groups based on students' abilities, perseverance, or work habits.

2. **PRODUCTIVITY:** A teacher may arrange groups that lead to the completion of more work. This may, for example, require there to be a strong leader in a group for project work, or the teacher may prefer to group weaker students with stronger ones.

3. **PEACEFULNESS**: A teacher may create groups that intentionally keep friends or disruptive students apart, as such groupings may lead to less productivity.

Alternatively, a teacher may group for *social goals* for several reasons:

1. **DIVERSITY**: A teacher may want to arrange students into groups so as to ensure a specific diversity (e.g., gender) exists within each group.

2. **INTEGRATION**: A teacher may use this method to push students out of their social comfort zones and help them collaborate with students with whom they don't normally associate.

3. **SOCIALIZATION**: Occasionally, a teacher may specifically want to reward performance or positive behavior by allowing students to work with their friends.

In high schools, on the other hand, we most often saw teachers allowing their students to form *self-selected* groups either for the whole lesson period or for a specific activity within a lesson. Although students may group themselves for the educational reasons listed (Cobb, Wood, Yackel, & McNeal, 1992; Yackel & Cobb, 1996), more commonly, students group themselves for social reasons (Urdan & Maehr, 1995). In a study where I looked specifically at the goals for how students self-select their groups, 95% of students either grouped themselves, or attempted to group themselves, in order to socialize.

THE PROBLEM

Regardless of whether teachers group for educational reasons or social reasons, there is almost always a mismatch between the teacher's goals and the students' goals (Kotsopoulos, 2007). For example, whereas a teacher may have pedagogical reasons for wanting students to work together, the students—wishing instead to work with their friends—may begrudgingly work in their assigned groups in ways that cannot be considered collaborative (Clarke & Xu, 2008; Esmonde, 2009). These sorts of mismatches arise from the tension between the individual goals of students concerned with themselves, or their cadre of friends, and the classroom goals set by the teacher for everyone in the room. Couple this with the social dynamics often present in classrooms, and a teacher faces a situation where students not only wish to be with certain classmates, but also disdain to be with others. In essence, no matter how strategic a teacher is in their groupings, when there is a mismatch between their goals and students' individual goals, it means some students will be unhappy and will disengage. This disengagement is antithetical to a thinking classroom. I would venture to say that most teachers are familiar with this challenge.

> No matter how strategic a teacher is in their groupings, when there is a mismatch between their goals and students' individual goals, it means some students will be unhappy and will disengage.

In the context of a thinking classroom, there is an even more problematic situation that arises out of both strategic and self-selected grouping strategies. It turns out that regardless of what strategy you use, the students know what their role in the group will be that day. And rarely is that role to think. To better understand this, I spent over 40 hours in classrooms where teachers were using one of these two grouping strategies observing and talking to students. In an overwhelming number of cases, whether strategically grouped or self-selected, 80% of students entered their groups feeling like they were going to be a follower rather than a leader—to be a follower rather than a thinker.

Interviewer	So, now that you know what group you will be in, what do you think your role will be?
Stuart	What do you mean?
Interviewer	I mean, are you going to offer any ideas or take a lead on any of the work?
Stuart	Probably not. I'm with Gabriel and Aisha and they are both brainiacs. That's why the teacher put me with them.
Interviewer	I notice that you are sitting with Francis, Nahal, and Deja again.
Amanda	Yup. I always do.
Interviewer	Your friends?
Amanda	Yup.
Interviewer	So, are you going to solve the problem today?
Amanda	No way. That's Deja's job. I'll just follow along.

These sorts of comments were so prevalent in my conversations with students that I ran a questionnaire on over 200 students in classrooms where teachers were using either strategic grouping or self-selected groupings. I asked two questions:

1. If I told you that next class you are going to work in groups to solve a word problem, what is the likelihood that you would offer an idea?

2. If you were to offer an idea, what is the likelihood that your idea would contribute to the solution of the problem?

> Students, by and large, know why they are being placed with certain other students, and they live down to these expectations.

For Question 1, over 80% of the students said that it was unlikely or highly unlikely that they would offer an idea, and over 90% said it would be highly unlikely that one of their ideas would contribute to a solution. Irrespective of the grouping method being used, the vast majority of students do not enter their groups thinking they are going to make a significant, if any, contribution to their group. They are entering the groups in the role of follower, expecting not to think. That means that with the strategic groupings, other than those 10% to 20% who are accustomed to taking the lead, the rest of the students, by and large, know why they are being placed with certain other students, and they live down to these expectations. Likewise, in self-selected groups, students fall immediately into the dictated patterns of leaders and followers that already exist within the social dynamics of children's lives. Either way, these patterns of behavior are antithetical to the goals of a thinking classroom.

To counteract this, some teachers have adopted the practice of assigning a role to each student in a group—leader, recorder, timekeeper, resource-getter, encourager, et cetera. But this assignment of roles only serves to exacerbate the aforementioned problem. I spent time in three classrooms where this practice was being used, and I observed even less authentic engagement with the activities. Rather than thinking about the task at hand, students were trying to adhere to their roles, which many found helpful in escaping from the actual work at hand—to solve a problem—to think.

While we clearly know the value and importance of student collaboration, the problem is that efforts to actualize collaboration through either strategic groups or assigned roles may be having a negative effect on how our students engage with each other and the task at hand. At the same time, self-selected groups seem to be of little help, as students group themselves for reasons antithetical to solving mathematical problems. You may have seen this, and been frustrated by it, in your own classroom as well. So, how to fix it?

TOWARD A THINKING CLASSROOM

Adhering to the contrarian approach of beginning with a practice that is opposite to what the norm is, we decided to try random groupings. Random grouping, from the perspective of nexus of control, was opposite to both strategic and self-selected grouping. This idea was implemented with 11 different teachers. Some were hesitant at first, but because they saw the same problems with the current methods they were using, they were willing to try. Our first iterations involved the teachers making new randomly assigned seating charts for each of their classrooms and then presenting these to their students with the explanation that these were random assignments.

This proved to be spectacularly ineffective. We were seeing no greater benefits with this method than with strategic groupings. Students were falling into preconceived roles and not engaging in ways reported in the literature on effective collaboration. Interviews with students immediately revealed the problem.

Interviewer	So, how do you like your new group?
Mitchel	It's OK I guess.
Interviewer	Did you like that the teacher picked the group randomly?
Mitchel	Yeah right!
Interviewer	You don't believe they are random?
Mitchel	Of course not.

Although the new groupings were random, the students did not perceive the randomness in it. Why would they? The teacher putting up a new seating plan and saying it is random is substantively not that different from just putting up a new seating plan—something they have experienced many times before and that they knew to be strategic. Although we had removed the nexus of control from the teachers, the students did not perceive this to be the case.

> The teacher putting up a new seating plan and saying it is random is substantively not that different from just putting up a new seating plan.

So, we did an immediate adjustment, and used playing cards to assign the groups. This was simply done by labeling each table (or desk) group with a card rank (2, 7, jack, queen, etc.) and having students draw a card from a deck to determine what group they would be in and where they would sit. This simple change had an overwhelming effect on the students' perception of how the groups were formed and where the nexus of control rested.

Interviewer	So, how do you like your new group?
Mitchel	Yup.
Interviewer	Did you like that it was created randomly?
Mitchel	Yeah. It was cool to pick a card.
Interviewer	Cool?
Mitchel	Yeah. I was hoping I would get a seven. Luis had already picked a seven.
Interviewer	What did you get?
Mitchel	A jack.
Interviewer	Hmm . . .
Mitchel	Maybe next time.

Although randomizing wrested the control from the teachers, making it *visibly random* was necessary for the students to both perceive and believe the randomness.

Although randomizing wrested the control from the teachers, making it *visibly random* was necessary for the students to both perceive and believe the randomness. We needed visibly random groupings. After several more iterations, with these 11 teachers and others, more nuances emerged. First, we learned that the randomization needed to be frequent—approximately every hour. If we left it longer, then we began to see the roles within the group calcify into their active and passive states. We needed frequent visibly random groupings.

We also learned that, from Grade 3 up, the optimal group size was three. Groups of two struggled more than groups of three, and groups of four almost always devolved into a group of three plus one, or two groups of two. This is because for a group to be generative, it needs to have both redundancy and diversity (Davis & Simmt, 2003). Redundancy, in this context, reflects things that a group of students has in common—language, interests, experiences, knowledge. Without these commonalities they cannot even begin to collaborate. But if all they have is redundancy, they will not achieve anything beyond what they enter the group with. To be generative, they also need diversity; the things that individual members of the group bring that are not shared by the others—different ideas, viewpoints, perspectives, representations, et cetera. Groups of three seem to have the perfect balance of redundancy and diversity. This is why self-selected groups tend not to be as productive—too much redundancy, not enough diversity.

Figure 2.1 A group of three students working collaboratively.
Source: SDI Productions /iStock.com

For Grades K–2, however, the optimal group size was two. Despite the lack of diversity this affords, students at this age are still developmentally in a stage of parallel play, and collaboration consists mostly of polite turn taking. What we learned was that groups of two, coupled with the guidance of the teacher, allows this polite turn taking to start to shift toward listening to each other and building on each other's ideas—to shift toward true collaboration. This is not to say that your eighth graders are demonstrating great collaboration in your classroom, but rather that they have the skills in place to do so, and often use these skills outside the classroom.

Once we were implementing frequent and visibly random groupings, we saw an immediate uptick in the amount of students' engagement and thinking. By removing the nexus of control from both the teacher and the students, the students entered their groups not knowing what their role would be that day. This allowed for different students to step forward and begin to think. We ran the aforementioned survey after two weeks of implementation, and we saw a definite increase in the number of students who would offer an idea. And after six weeks almost 100% of students said that they were either likely or very likely to offer an idea. This, despite the fact that only 50% believed that their idea would lead to a solution. The students were willing to try, irrespective of whether their idea would lead to a solution.

Aside from an increase in thinking we saw several other benefits in our implementation of frequent visibly random groups.

Willingness to Collaborate

Although many students rail against the groups they find themselves in on Day 1, at the three-week point this resistance is usually completely gone, and they are open to working with anyone they are placed with.

Researcher	So, I noticed that last week you tried a few times to sit with Jackson. Are you still trying to do so?
Hunter	No.
Researcher	Why not?
Hunter	At first, I thought that the teacher was trying to keep us apart. Then, on Friday, we got to work together.
Researcher	So, do you still think the teacher is trying to keep you apart?
Hunter	No. I don't think she likes us working together, but when the cards came up the way they did, she didn't change it. I guess it's up to the cards now.

Researcher	I saw what you did last week.
Jasmine	What do you mean?
Researcher	I saw how you switched groups.
Jasmine	Oh that. That's nothing.
Researcher	But you didn't do it this week. Why not?
Jasmine	I guess it doesn't matter so much who you are working with. I mean, it is just for one class, and then the groups change again.

Elimination of Social Barriers

When teachers allow students to self-select, what we see is often a reflection of the social structures easily observable in the hallways.

Students choose their friends, their affinity groups, or their social groups. These social structures can create barriers to collaboration in the classroom. With visibly random grouping, these barriers begin to fall away.

Researcher   Tell me about how your group work went today.

Melanie   Fine.

Researcher   Who were you with?

Melanie   I was with Aisha and Luis.

Researcher   Can you tell me a little bit about Aisha or Luis?

Melanie   Ok. Aisha is smart. I worked with her one time before. She really knows what is going on, so I try to listen carefully to her when she has something to say. She's in my science class as well, and her sister is in my English class.

Researcher   How do you know that Aisha's sister is in your English class?

Melanie   We figured it out today.

When students worked with new random partners every hour, they began to cross social boundaries and form an awareness about each other in ways that were not happening before.

Increased Knowledge Mobility

As mentioned, when students work in self-selected groups, the social barriers stay intact. These social barriers, in turn, create barriers to knowledge moving between groups. When the social barriers come down, so too do the barriers to knowledge mobility.

Researcher   Good problem today, huh? Can you tell me how you guys solved it?

Idris   Yeah, that was a tough one. We were stuck for a long time.

Researcher   What did you eventually figure out?

Idris   Well, we saw that the group next to us was using a table to check out some possibilities, and we could see that there was a pattern in the numbers they were using, so we tried that. That sort of got us going and we got an answer pretty quickly after that.

Researcher Was it the right answer?

Idris It was, but we weren't so sure. The group next to us had a different answer, and it took a long time working with them before we figured out which one was correct.

Knowledge mobility takes one of three forms: (1) members of a group going out to other groups to *borrow an idea* to bring back to their group, (2) members of a group going out to compare their answer to other answers, or (3) two (or more) groups coming together to debate different solutions. Or, like it did for Idris's group, it takes on a combination of these forms.

Increases in knowledge mobility were accompanied by a decrease in groups' reliance on the teacher and an increase in reliance on themselves (intragroup reliance) and other groups (intergroup reliance). The teacher was no longer the only source of knowledge in the room.

Increased Enthusiasm for Mathematics Learning

With the elimination of social barriers, students begin to enjoy mathematics class more.

James Math is now my favorite subject.

Jasmine I love this class. I mean, math isn't my favorite subject. But I love coming here.

Kendra And the beginning of every class is a bit of an adventure when we get to find out who we work with.

Along with this greater enthusiasm we also saw decreases in student absences and lateness.

Reduced Social Stress

Despite potential early resistance to visibly random grouping, once it is up and running, many students come to enjoy the elimination of the social stress involved in self-selecting groups. The students who most benefit from this are the students who classify themselves as shy.

Researcher	How are you liking the random groups?
Amanda	Love it. I don't care who I'm with as long as I don't have to try to get into a group myself.
Researcher	Why is that?
Amanda	I'm shy. In social studies the teacher always makes us pick our own groups. I hate that. I hate that feeling of asking if I can join a group and then being told no. I just want to work by myself in that class. But I don't want to work by myself. It's just so hard.

This relief is not restricted to only those who do not have strong social bonds in the classroom.

Interviewer	How are you liking the random groups?
Le	It's good. I like that I don't have to pick my groups.
Interviewer	Really! I thought you always wanted to be with the other two girls, . . . umm . . .
Le	Jennifer and Hillary.
Interviewer	Right. Aren't you friends any more?
Le	Oh yeah. We're friends. But I don't always want to work with them. We never get anything done when we're together. And it's tough to say that to them.
Interviewer	So . . .
Le	Random groups helps me with that.

This relief is so profound that on several occasions I have seen students reminding the teacher to put them into visibly random groups.

Q In this chapter you talk about mobility of knowledge as a good thing. Won't the students just come to rely on this and, instead of doing their own thinking, they will just copy from other groups?

A No. In the hundreds of lessons I have observed where frequent visibly random groups were being used, I have seen fewer than 10 instances of a group copying from another group. Students tend not to treat knowledge mobility—students call it *borrowing an idea*—as a way to reduce thinking. Rather, they use it as a way to keep thinking when they are stuck.

Q Will the students whose idea is being borrowed be OK with it?

A The answer depends on the current culture in your class. In settings where students are used to competing for praise and prestige, groups will likely try to hide their answers from others. Thanks to the ubiquity of visibly random groups, however, and the subsequent elimination of social barriers, group boundaries begin to become porous—although group boundaries are defined for the period, these boundaries are clearly temporary and arbitrary. This allows them to also be seen as open, and, with the free movement of members from one group to another, ideas naturally begin to flow across boundaries.

Q You say that students will become agreeable to work in any group they are placed with, but I can think of some students that will not like this. Will they eventually?

A Students being agreeable to work in any group they are placed in is not the same thing as students liking frequent visibly random groups. You will have students who would still prefer to work alone or pick their own groups. But our research shows that even these students do more thinking when in random groups.

Q You make it sound like mobility of knowledge will automatically start to happen. I have now been doing random groups for a while, and I'm not seeing it. Is there anything I can do to help it along?

A I will discuss this at length in Chapter 8 through the lens of fostering student autonomy. For now, however, you can help it along by pointing out to groups where, in the room, they can find the knowledge they might need. You can also put groups together that you know need to move some knowledge between them.

Q When I use a deck of cards to create random groups, I get some groups of four and some groups of two, and I always seem to have one student who has no partners. How do you avoid this?

A You need to set up the deck before you begin. First of all, have only three of each rank of card in your deck—three aces, three tens, et cetera. If you're teaching Grades K–2, make this two of each rank. Second, make sure the total number of cards in your deck is equal to the number of students enrolled in your class. Finally, if some students are absent, remove a single card from each rank until you have the correct number of cards. This will ensure that there will never be a group of size one. It also has the benefit that if a student arrives late, you simply let them pick a card from the remaining singletons and plug them into an already functioning group of two. This avoids having to make a new group out of students who arrive late.

Q I have some students who are constantly switching cards, or not going to the group that the card specifies. How do I deal with that?

A There are two ways to deal with this. First, you can target that student and ask them to show you the card after they have picked it. Second, you can have all students show their card after they have chosen it. Both of these methods create the impression that you know what card a specific student has and that you will be watching to make sure they go where they are supposed to. It sounds crazy, but the students really believe that you know all their cards.

Q If I don't want to use cards, are there other methods to randomize groups, and are they as good?

A Some teachers like to use popsicle sticks with student names, name cards created by the students, or photographs of students to make groups. Others like to use technology. It doesn't matter what you use, as long as it is random, and the students perceive it as random. Technology tools tend to be a bit of a black box, so sometimes students don't trust that they are being truly random. To enhance the perception of randomness, you can let one student roll a dice and then come up and push the button on a digital randomizer as many times as the dice shows.

> Transitions are more easily facilitated if your method of randomization doesn't just tell them what group they are in, but it also tells them where to go to meet their group.

Regardless of what method you use, however, we found that transitions are more easily facilitated if your method of randomization doesn't just tell them what group they are in, but it also tells them where to go to meet their group. This is most easily accomplished by labeling clusters of desks or sections of the vertical boards (see Chapter 3) with the same labels as your randomizer. So, if you use cards, label the workstations with A, 2, 3, . . ., J, Q, K. If you use a digital randomizer that names the team by letters or colors, then use these to label the workstations.

Q I teach my kids all day and want to have them in groups for lots of different subjects. Should I be randomizing them for every class?

A Although randomizing the groups about once an hour proved to be best, this is not practical if you have your students all day.

In those cases, we made use of natural transitions like lunch, recess, and coming back from the library or gym. In essence, you randomize them every time they come back into the classroom.

Q I have my students in desks. And they have all their stuff in their desks. Trying to move these around several times a day will be a nightmare. Any suggestions?

A Only in elementary schools do we allow students to own real estate in the classroom. And with this comes a heightened sense of entitlement, ownership, and protectiveness over their real estate. You may have seen students getting stressed when someone sits in *their* desk. We found that the best way to deal with this is to get their stuff out of the desks and store it in bins elsewhere in the classroom.

Q I can think of two students right now who absolutely should not be together. How do I deal with this?

A There are several reasons we may wish to keep students apart from each other. Ironically, regardless of the reason, those students often want to be together. And that will happen when you start randomizing your groups—usually on the first day. The best thing to do when this happens is to visit that group first and just say, "Are we going to have a problem here?" More often than not, these students are so thrilled to be together that they do not want to ruin their chances of it happening again.

If the situation is one where the students don't want to be together, then the intervention is different. Treat them being together like it is the most normal thing in the world—just like any other group. The more normal you make it, the more likely they are to be OK with it. Students are highly attuned to the things that you worry about—if you worry about them being together, then they will fixate on that as well. If, despite this, you sense trouble brewing, visit the group, and remind them that it is just for one class and that you expect that they will be respectful to each other.

If the situation is one where the students really shouldn't be together, then you will find a way to work around this. However, be certain that it is the case that they *really shouldn't* be together and not that you perceive or prefer that they shouldn't be together. There is a difference.

Q I'm worried what will happen when a weaker student ends up in the same group with a stronger student. Won't that weaker student get excluded or marginalized?

A I will discuss this issue in greater detail in Chapter 9. For now, however, I will just say that aside from mobilizing knowledge, frequent visibly random groups also mobilizes empathy. As a society we give far too little credit for the empathy that children have for each other. Perhaps this is a by-product of our efforts to stem school and social media bullying and exclusion. In our efforts to stem these social ails we build up an assumption that, without our interventions, all kids would be capable of perpetrating such acts. But this is not true. Children have an unbelievable capacity for empathy for each other. They know which students are strong and weak, which have special designations, and which receive adaptations or accommodations, and yet they still see each other as peers and as friends. Random groups puts this empathy into motion and gives it a venue to play out in. I have many times been in classrooms where I have observed students make room in their group for someone who is academically weaker. For example, the following exchange took place in a Grade 10 classroom.

> **Phil** OK, Amber, you are going to be our calculator girl. Whenever there is something to calculate, you are going to help us with that.
>
> **Steve** You. You're the one—the one that is going to help us get through this.

Amber is a girl with an intellectual disability, and these boys found something that Amber could be successful with, supported her in this, and celebrated her and with her when she "helped them get through it."

Q For a long time now I have been working on differentiating my instruction for my learners. This works best if my students are in ability groupings. How will that work if I start to randomize them?

A It will be difficult for you to believe this right now, but over time it will be fine. This is for three reasons. First, with the integration of other thinking practices from this book, *all* your students will become better at thinking—and they will do so in ways that are not predicted by your current perception of your students' abilities. Some of your strong students will reveal themselves to only be strong at mimicking and will struggle on thinking tasks. Some of your weaker students will prove themselves to be much better at thinking than others in the class. And the ones in the middle will completely reshuffle your perceived hierarchy. So, whatever ability groupings you thought you would make would not be homogeneous.

Second, differentiation looks different in a thinking classroom. I will discuss this in detail in Chapter 9. For now, however, I will just say that in thinking classrooms we start all groups on the same task and then differentiate the hints and extensions we give each group depending on how they are doing.

The third reason not to worry about this is that, as mentioned earlier, diversity is a strength—diverse groups are able to be more creative and more generative.

Q Sometimes I like to let students think and work on a task before they go into their groups. Should I still do that with random groups?

A No. In our research we found that when we did this, it created too much diversity. The students who knew what they were doing would complete the task, while the ones that had no clue made no progress. When these students came together in random groups, the differences in their abilities was then too great, and less collaboration took place.

SUMMARY

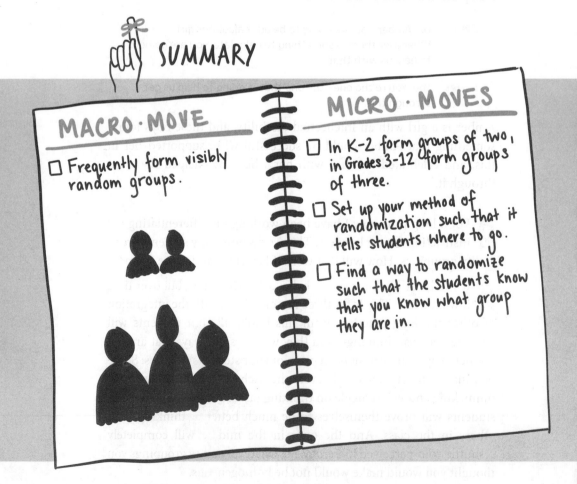

MACRO · MOVE

☐ Frequently form visibly random groups.

MICRO · MOVES

☐ In K-2 form groups of two, in Grades 3-12 form groups of three.

☐ Set up your method of randomization such that it tells students where to go.

☐ Find a way to randomize such that the students know that you know what group they are in.

QUESTIONS TO THINK ABOUT

1. What are some of the things in this chapter that immediately feel correct?

2. What is the worst combination of students that can come together in random groups? What is this perception based on? Is there any possibility that this could go well?

3. Can you think of some students who would benefit greatly from visibly random groups? Why would they benefit?

4. Can you think of some students who will likely not enjoy random groups? Would it still be good for them?

5. What are some of the challenges you anticipate you will experience in implementing the strategies suggested in this chapter? What are some of the ways to overcome these?

☑ TRY THIS

As mentioned in the introduction and Chapter 1, the practices in the first three chapters are best implemented together. In preparation for this, figure out what your preferred method of randomizing the students is. Keep in mind that whatever method you choose should not just tell your students what group they are in, but also where to go.

CHAPTER 3
WHERE STUDENTS WORK IN A THINKING CLASSROOM

Now that we have discussed thinking tasks and have put our students into visibly random groups, we need to find somewhere for them to work. The norm is to have them sit together at a table or a cluster of desks and to do their work in their notebooks. But is this the most conducive workspace if we're trying to build a thinking classroom? This chapter presents the results of explorations around alternate workspaces and the impact that they have on creating and sustaining thinking. By the end of the chapter you will learn what the optimal workspace is for thinking as well as what it is about other workspaces, including desks and notebooks, that makes them so ineffective.

THE ISSUE

One of the most enduring institutional norms that exists in mathematics classrooms is students sitting at their desks (or tables) and writing in notebooks. With the exception of some primary classrooms, I saw this in every classroom I visited. Students sit and take notes in their notebooks. They sit and do now-you-try-one tasks in their notebooks. And they sit and do their homework in their notebooks. In fact, there are some students who spend more hours of their day sitting and writing in their notebooks than they spend sleeping. So, should we add to this by also having students sit and work in their notebooks while working on thinking tasks?

THE PROBLEM

That would be fine if this were the workspace most conducive to initiating and sustaining thinking. The problem is that it is not. As mentioned, the notebook has become the catchall for all student work in the mathematics classroom. From taking notes to completing now-you-try-one tasks to doing homework, the notebook is where students do their work. All these activities, in and of themselves, are very different from each other. Yet, because they are all done by sitting and writing in notebooks, the students bring the same behavior and level of energy, engagement, and attention to all three activities. Note taking, as we will see in Chapter 10, is a largely passive activity, which, as we saw in Chapter 1, promotes mimicking. As we saw in the introduction and will see in Chapter 7, both now-you-try-one tasks and homework also rely heavily on mimicking. In short, sitting and working in notebooks promotes and rewards,

> The continuity of the workspace ensured a continuity of student behaviors.

in continuous and ubiquitous ways, passive mimicking behaviors. If we then ask students to work their way through thinking tasks while sitting and working in their notebooks, it is not surprising that we get the kind of result I did when visiting Jane's class all those years ago. The continuity of the workspace ensured a continuity of student behaviors. And when those behaviors did not produce results, students quickly gave up.

> Just because sitting and writing in the notebook is the obvious place for *some* activities, it does not have to be the workspace for *all* activities.

So, what do we do? Students need a place to record things for future reference. They need a place to do their homework. The notebook is the obvious place to do that. I don't disagree with that, and I will deal with these topics in Chapters 10 and 7, respectively. But just because sitting and writing in the notebook is the obvious place for *some* activities, it does not have to be the workspace for *all* activities.

TOWARD A THINKING CLASSROOM

We began experimenting with alternative workspaces for the part of the lesson where students are working on thinking tasks. Adhering to my contrarian design principles, the first thing we tried was having students stand and write on portable whiteboards that were lying on their desks. This produced positive results in classrooms where we could remove the chairs, but in situations where the desks were attached to the chair, students found their way back into their seats. Although this still produced more thinking and engagement than having them write in their notebooks, we still saw non-thinking behaviors like stalling and faking.

We tried having the students standing at wall-mounted whiteboards when working through the thinking tasks (see Figure 3.1). This almost completely eliminated the stalling and faking behavior and had a huge effect on the amount of time students were willing to spend thinking when working on thinking tasks. When we coupled this with the benefits of visibly random groups (Chapter 2), the thinking behavior increased by an order of magnitude. In the 15 years that I have been engaged in the thinking classroom research, nothing we have tried has had such a positive and profound effect on student thinking as having them work in random groups at vertical whiteboards. Students were thinking longer, discussing more mathematics, and persisting when the tasks were hard.

Figure 3.1 Students engage with wall-mounted whiteboards.

The changes were so transformative that I decided to run a controlled experiment wherein we compared student thinking and engagement for a number of variables across five different workspaces—students standing at vertical whiteboards, sitting at horizontal whiteboards, standing at flipchart paper mounted on the wall, sitting at flipchart paper on a table, and sitting with their notebooks. Students were put into groups and randomly assigned to one of these five workspaces. In each classroom that we ran the experiment, all five workspaces were being used.

Once assigned to a workspace, all groups in that classroom were given the same thinking task to work on while we timed their behaviors across three variables:

1. how long (in seconds) it took them to start talking about the problem,

2. how long (in seconds) it took them to make their first mathematical notation on whatever workspace they were assigned to, and

3. how long (in minutes) they were willing to keep working without the teacher needing to encourage them to keep going.

We also used a scale of 0 (none) to 3 (lots) to score their behaviors on a further six variables:

4. how much discussion there was,

5. how eager and enthusiastic they were to start,

6. the degree to which every member of the group participated in the activity,

7. how persistent they were at trying to solve the problem,

8. the amount of knowledge mobility, and

9. the degree to which their written work was non-linear.

Regarding Variable 8, our initial attempts at having students work on vertical whiteboards had shown us how much movement of ideas occurred and how positive it was, and I wanted to capture this in the data. With Variable 9, my studenting research had shown that when students are not mimicking, their work tends to be messier than when they are mimicking, and although this is not a perfect indicator of thinking, I wanted to capture it in the data as well.

This experiment ran in five different classrooms. Figure 3.2 shows the average time and scores for a given workspace and a given variable across these five different classrooms.

WORK SURFACE	vertical whiteboard	horizontal whiteboard	vertical paper	horizontal paper	notebook
NUMBER OF GROUPS	10	10	9	9	8
1. time to task (seconds)	12.8	13.2	12.1	14.1	13.0
2. time to first notation (seconds)	20.3	23.5	144.3	126.8	18.2
3. time on task (minutes)	7.1	4.6	3.0	3.1	3.4
4. eagerness to start	3.0	2.3	1.2	1.0	0.9
5. amount of discussion	2.8	2.2	1.5	1.1	0.6
6. amount of participating	2.8	2.1	1.8	1.6	0.9
7. amount of persistence	2.6	2.6	1.8	1.9	1.9
8. amount of knowledge mobility	2.5	1.2	2.0	1.3	1.2
9. non-linearity of work	2.7	2.9	1.0	1.1	0.8

Figure 3.2 Average times and scores on the nine measures.

These results clearly show that having students work on whiteboards produced better results across almost all variables than if the students worked on flipchart paper—irrespective of whether they were standing or sitting. The data also show that across almost all variables, any alternate workspace produced better results than having students work through thinking tasks in their notebooks while sitting at their desks. Observations and interviews with students during this experiment and in other settings have since revealed that the main reason for these results is that when they work on whiteboards, they can quickly erase any errors, which, for them, reduces the risk of trying something. While the private, familiar, and small nature of notebooks provides this same sense of low risk, the very public and permanent nature of the flipchart paper creates more risk and places greater emphasis on correct answers. The time to first notation across these five workspaces is reflective of these insights from students.

> When students work on whiteboards, they can quickly erase any errors, which, for them, reduces the risk of trying something.

The results also show that, with both flipchart paper and whiteboards, standing produced better results than sitting. This is not surprising. We know from physiology that standing is better than sitting—and not just in a *sitting is the new smoking* kind of way. Standing necessitates a better posture, which has been linked to improvements in mood and increases in energy (Peper & Lin, 2012; Wilson & Peper, 2004). We also know that the majority of communication is non-verbal (Mehrabian, 2009), consisting of gestures, facial expressions, tone of voice, and body language. Standing gives a larger canvas for these forms of non-verbal communication (Wells, 2014).

Standing also afforded an increase in knowledge mobility. Having students work vertically makes their work visible to everyone in the room, thereby increasing the porosity between groups, which, in turn, heightens the possibility that ideas will move between groups. As mentioned in Chapter 2, increases in knowledge mobility also increased students' reliance on each other—both within and between groups—while at the same time decreasing their reliance on the teacher as the only source of knowledge in the room.

However, all of these reasons for why vertical work surfaces produce better results are trumped by something that emerged out of conversations with students—slowly over many years. It turns out that when students are sitting, they feel anonymous. And the further they sit from the teacher and the more things—desks, other students, computers, et cetera—are between them and the teacher, the more anonymous

> When students are sitting, they feel anonymous. And when students feel anonymous, they are more likely to disengage.

they feel. And when students feel anonymous, they are more likely to disengage—in both conscious and unconscious ways. When students feel anonymous, they are consciously aware that they can shift their focus from the tasks at hand. What they shift this focus toward can range from completing homework for another class to playing on their phones. Regardless, this is a conscious decision, and they are aware and deliberate about where they choose to put their focus. But this shift in focus can also be unconscious, away from the task at hand. What that focus shifts to is often a sort of passive and automatic checking for social media notification on their phone—something students do often when they feel bored. But it can also consist of just *drifting away, checking out,* or *zoning out,* as was observed frequently by us and labeled by students upon seeing pictures of themselves.

> Having students standing immediately takes away that sense of anonymity and, with it, the conscious and unconscious pull away from the tasks at hand.

Having students standing immediately takes away that sense of anonymity and, with it, the conscious and unconscious pull away from the tasks at hand. This is not to say that students feel outed or on display in any sort of way. When all the students are working on vertical whiteboards, they do not feel unsafe. They just don't feel safe to get off task. This, coupled with the non-permanence afforded by the whiteboards, made the vertical whiteboards the best workspace for students to do their thinking.

Figure 3.3 Student groups utilize non-permanent surfaces.
Source: Photo courtesy of Alex Overwijk. Used with permission.

At the same time, vertical whiteboards offered an extra advantage from the teachers' point of view. From our own observations during this experiment, as well as in numerous settings using vertical whiteboards since then, we know that this workspace provides teachers with an ability to see everything that is happening in the room, and this enhances their ability to know at all times where a group's thinking is, how far they have progressed on the task, and when and where it's necessary to provide hints and extensions—something that will be discussed at length in Chapter 9. In other words, it aids teachers in their continual formative assessment and ability to provide and solicit feedback.

Most teachers, however, do not work in classrooms with whiteboards to accommodate 8–12 groups. For the last 20 years, classrooms have been outfitted with fewer and fewer whiteboards and more and more technology. Brand new schools often provide teachers with only one small whiteboard on the assumption that their teaching practice will primarily utilize technological alternatives to writing on boards. Retrofitting these classrooms with enough whiteboards for eight or more groups to work comfortably can be expensive. Fortunately, there are a number of very good, and inexpensive, alternatives. For one, blackboards work just as well. Windows also work as vertical erasable surfaces, as do vinyl picnic table covers, shower curtains, and cellophane (the kind you make gift baskets out of or wrap flowers in). For a more rigid alternative there are products that can be bought at home improvement stores. These are made of medium-density fiberboard with a white melamine finish on one side and are often referred to as shower board. At the same time there are a number of commercial products such as Wipeboards by Wipebooks, Better than Paper by Teacher Created Resources, and Dry Erase Surface by Post-It that have been manufactured specifically to function as whiteboard alternatives. With this wide range of alternatives available, there has never been a classroom that we could not outfit, in some way, to accommodate every group working on a vertical non-permanent surface. And because of all these alternatives, we stopped talking specifically about whiteboards and instead started referring to them as vertical non-permanent surfaces (VNPSs) after the qualities that proved to make whiteboards so conducive to thinking—vertical and easily erasable.

The aforementioned control experiment aside, from the moment we first tried having students work on VNPSs everything we have seen indicates that this is an effective way to increase student thinking and engagement. When coupled with random groups, non-thinking behaviors like slacking, stalling, and faking, for the most part, fall away. When coupled with the use of thinking tasks given early in the lesson the ability to mimic disappears. What is left is an environment that not only supports thinking, but also necessitates it. Since I began this research over 15 years ago, and ever since, I have never found a workspace that even comes close to these results.

> What is left is an environment that not only supports thinking, but also necessitates it.

This is not to say that the effectiveness of VNPSs cannot be enhanced by micro-moves, and, just as they did with the research on random groups, a number of these emerged and were experimented with during our research. For example, it works better if groups are close to each other without being crowded. The proximity of groups has a big impact on how well knowledge moves between groups. Likewise, using vertical whiteboards is enhanced by each group having only one marker. When every member of the group has their own marker, the group quickly devolves into three individuals working in parallel rather than collaborating. Finally, the experience is made easier for you as a teacher if you carry a marker that is a different color than the others in the room. This allows you to quickly discern where you have contributed and what you contributed last time you visited a group. And if your color is prominent and consistent, neighboring groups start to attend to the hints and extensions that you have left behind and use that information to keep themselves moving forward.

 FAQ

Q You again talk about knowledge mobility as if it is a good thing. Won't the vertical surfaces just make it easier for groups to copy each other?

A Again, in all the years I have been in classrooms where teachers are using VNPSs, I can count on one hand the number of times that I saw a group copying line for line, symbol for symbol, what another group had done, without doing any thinking on their own. It just doesn't happen. In a culture that values thinking—as opposed to answers—there is no motivation to just get the answer. It's the thinking that matters. So, the students may look around and use what others have done as inspiration for what they should try next—sometimes even talking to other groups about what they have done. While working on thinking tasks, students generate lots of ideas to try. Knowledge mobility is just another source for ideas for them to try and is a natural consequence of the porosity of group boundaries described in Chapter 2.

> In a culture that values thinking—as opposed to answers—there is no motivation to just get the answer. It's the thinking that matters.

Q If we are only giving each group one marker, how do I ensure that everyone is contributing?

A The quick answer to this is that you move the marker around. This can be done using varying degrees of subtlety. For example, every time you visit a group you casually ask for the marker from whoever is holding it, and when you leave you hand it to a different member of the group. When you are first starting out, giving it to the student standing furthest from the board is a good strategy. This is subtle. Less subtle is asking the group who hasn't held the marker yet and giving it to that student. Even less subtle is setting a timer that is loud enough for all students to hear and telling them that every time the timer goes off, they must pass the marker to another member of the group. After a very few weeks, students will begin to automatically move the markers around.

Q Moving the marker sounds great, but what if I have a student who has nothing to contribute. What good is it going to do them to have the marker?

A In these cases, you can add a rule, either for just this group or for the whole class, that whoever is holding the marker is not allowed to write any of their own ideas—they can only act as a scribe for what others say. Not only does this improve communication in the group, but it ensures that the group moves at the pace of the slowest learner. It also reduces the likelihood that one very quick thinker takes over the work while the others watch passively. If you couple this with the strategy that a group does not get the next task or extension unless every member of the group can explain how they solved the previous task, then you are necessitating that group members take care of each other's learning.

Q My class has students with a wide range of abilities. How well will this work for them?

A First, every class has a wide range of abilities. This is a defining quality of classrooms. However, the data upon which you are basing this assertion come from how students performed with your previous classroom norms. Once you get students thinking in random groups and on vertical surfaces, the playing field is sufficiently altered to allow new abilities to emerge. Every teacher who has done this has come to the realization that some of who they thought were their best students are actually quite weak at thinking tasks, and some of who they thought were their weak students are actually very good at thinking tasks.

This is not to say that your previous information is completely wrong, but rather that it is not as relevant when you go random and vertical.

> When students get into their groups and start working on vertical surfaces, the skills they need to be successful are things like communication, perseverance, patience, self-reliance, et cetera.

The wide range you see in your students' abilities when they are working individually in their notebooks is a product of their hugely varied acquisition and retention of mathematical knowledge. When students get into their groups and start working on vertical surfaces, the skills they need to be successful are things like communication, perseverance, patience, self-reliance, et cetera. And although these skills will vary throughout the room, the variance is typically not as great as with mathematical knowledge.

Q I have tried this, but I find that my students are too eager to erase things that are wrong, and then they lose track of what they have done. How do I stop this?

A The freedom to erase is vital to the students feeling safe to try and fail and try again. So, at one level, we do not want to stifle their freedom. At another level, however, when they erase too much. it can become detrimental to their ability to move forward. At the same time, we want them to be comfortable with their mistakes. There are two answers to this problem. The first is to wait. The urge to erase typically goes away after a while, and they begin to erase only when they run out of room. The second answer is to talk to the students, but only after they have been working on VNPSs for a few weeks. You can suggest that they draw a box around things they want to erase and draw a single line through it—call it slow garbage. Alternatively, or additionally, you can talk about how all ideas are valuable, and even things that are wrong have some value. Either way, you need to be careful not to constrain their freedom to erase too much or too soon.

One area where you can impose more constraints is around the erasing of others' work. This can diminish a student's contribution, which is never OK. You should set rules that students are not allowed to erase someone's work without their permission.

 # SUMMARY

MACRO·MOVE

☐ Use vertical non·permanent surfaces (VNPSs).

Who in this group hasn't held the marker yet?

MICRO·MOVES

☐ Have only one marker per group.

☐ Move the marker around within the group.

☐ Sometimes have the rule that the person writing cannot write any of their own ideas.

☐ Hold groups responsible for the learning of every member of the group.

☐ Have groups in close (but not too close) proximity to each other.

☐ Talk to the students about valuing wrong ideas & not erasing others' work.

QUESTIONS TO THINK ABOUT

1. What are some of the things in this chapter that immediately feel correct?

2. What must change in your room in order for you to gain the wall space necessary to get all of your students working on a VNPS?

3. Which strategy for moving the marker around did you like the best and why?

4. In this chapter you read about the notebook as a catchall— the place where we default to having students do their work. Think about all the different types of things you ask students to do in their notebooks. Which of these, other than doing

thinking tasks, can you imagine having your students doing on VNPSs?

5. What else happens in your classroom that can be enhanced by having the students work on a VNPS?

6. What are some of the challenges you anticipate you will experience in implementing the strategies suggested in this chapter? What are some of the ways to overcome these?

☑ TRY THIS

Now that you have read the first three chapters, you may want to try some of these ideas out in your classroom. As mentioned in the introduction, it is best to implement the ideas of the first three chapters—thinking tasks, frequent visibly random groups, and VNPSs—together. The reasons for this will be discussed more in Chapter 15. For now, however, if you are ready to try these ideas, I provide some highly engaging non-curricular thinking tasks here for you to start out with. Recall from Chapter 1 that it is best to do three to five lessons of these types of tasks before shifting to scripted curriculum thinking tasks. And recall from Chapters 2 and 3 that it is important to have students in random groups working on VNPSs. Utilizing all of the micro-moves from each of the first three chapters will enhance the experience. To start with you might want to focus on these:

- Avoid setting up the tasks so they can be solved through mimicking.

- Randomize in a way that is visible to students.

- In Grades K–2 form groups of two, and in Grades 3–12 form groups of three.

- Set up your method of randomization such that it tells students where to go.

- Have only one marker per group.

- Have groups in close (but not too close) proximity to each other.

Grades K–3: What color am I?

What color is each shape if

- blue has no corners,
- green is between red and black,
- green is on the left of orange, and
- purple is next to red.

Source: Adapted from the Coloured Shapes task by © Crown Copyright 2000.

Grades 4–7: How many 7s?

If I were to write the numbers from 1 to 100, how many times would I use the digit 7? What if I wrote 1 to 1000? How any zeros?

$$1, 2, 3, 4, 5, 6, 7, 8, 9, 10, 11, \ldots, 997, 998, 999, 1\,000$$

Grades 8–12: Split 25

Decompose 25 using addition. For example,

$$25 = 10 + 15$$
$$25 = 10 + 10 + 5$$
$$25 = 3 + 3 + 3 + 16$$

What is the biggest product you can make if you multiply the addends together?

(Note: The examples bias an assumption that the addends must be whole numbers. However, the instructions do not specify this. Let this fact emerge after they have found a maximum for whole numbers.)

Source: Adapted from a task by Malcolm Swan.

CHAPTER 4

HOW WE ARRANGE THE FURNITURE IN A THINKING CLASSROOM

Sept. 1st

Welcome class!

A thinking classroom is defined, largely, by the kinds of activities that students engage in and how the teacher facilitates these activities. So far, I have discussed a number of results that show how thinking can be occasioned through the tasks we choose, how we group students, and where they work. These practices have been proven, over and over again, to increase the amount of time students spend thinking in the classroom. But what happens when the students and the teacher go home, and all that is left is the room and what is in it? Is it still a thinking classroom? Obviously not. If there are no students to do the thinking, it is not a thinking classroom. Or so I thought. In this chapter you will learn about the results of the research into how the physical organization of classroom furniture affects student thinking and how any classroom can be reorganized to help optimize student thinking.

> What happens when the students and the teacher go home, and all that is left is the room and what is in it? Is it still a thinking classroom?

THE ISSUE

At its core, a classroom is just a room with furniture. Absent the students and the teacher, a classroom is an inert space waiting to be inhabited, waiting to be used, waiting for thinking to happen. This is not to say that the classroom, in its inert form, has no role in what happens in it. It actually has a huge role in determining what kind of learning can take place in it. As an extreme example, a gymnasium allows for the possibility of a type of learning that is different from a woodwork shop, which, in itself, allows for a type of learning that is different from an art room or a music room. In short, classroom spaces are designed for specific types of learning. This is true of a mathematics classroom as well. Different classroom setups allow for different types of learning. We have already seen this with respect to vertical non-permanent surfaces. Classrooms with lots of VNPSs allow for the possibility of a different type of learning experience than classrooms without them. But how a classroom is set up goes well beyond whether it has whiteboards or not.

THE PROBLEM

Early on in the research into building thinking classrooms, I made an interesting observation. Every time we worked in classrooms that were super organized—desks or tables in perfect rows, in-baskets and out-baskets for all eventualities,

> **Every time we worked in classrooms that were super organized we had more difficulty generating thinking.**

everything color coded and in its place—we had more difficulty generating thinking. It didn't matter whether we were experimenting with vertical surfaces, random groups, how to answer questions, homework, or something else. If the room was super organized, we had more difficulty generating positive results. But, when we were working in classrooms that were disorderly, but not overly so, we had better results. What was it about those super organized classrooms that were negating some of our otherwise effective practices for generating thinking?

> **Thinking is messy. It requires a significant amount of risk taking, trial and error, and non-linear thinking. It turns out that in super organized classrooms, students don't feel safe to get messy in these ways.**

Thinking is messy. It requires a significant amount of risk taking, trial and error, and non-linear thinking. It turns out that in super organized classrooms, students don't feel safe to get messy in these ways. The message they are receiving is that learning needs to be orderly, structured, and precise. In these perfectly organized classrooms, the physical spaces in which they are being asked to think is incommensurate with the messiness of thinking. This is a problem. On the other hand, thinking should not be completely unstructured. It needs elements of representation and organization for patterns to begin to emerge. Therefore, overly chaotic spaces are not the answer either.

When I looked closely at the types of classrooms from which positive results were consistently emerging, it became clear that they were neither too organized nor too chaotic. They were relaxed spaces in which students felt safe to take risks, to try, and to fail. At the same time, they were not so chaotic that the physical structure of the classroom became a distraction to the students. It seemed that a classroom needed to have a just-right amount of disorder for thinking to flourish. I wanted to find this minimum amount of disorder.

 TOWARD A THINKING CLASSROOM

It turns out that how desks and tables are arranged within a classroom says more about what kind of learning behavior—and hence, thinking behavior—is expected in that room than anything else. A teacher may have a relaxed attitude about precision, but if the desks are in razor straight rows, that classroom is telling students that orderliness matters.

Think about the last time you attended a professional development session. If the room was full of chairs, had no tables, and had a podium, you knew you were going to get a lecture—and you knew this long before the session started. If you walked in and there were tables, neatly organized into rows with chairs only on one side so that everyone was facing the front, you knew you were also likely to get a lecture—but with the possibility of some activity. If there were tables

> When you walked into the room, you knew immediately what to expect—and that expectation shaped your behavior.

that had chairs all the way around them, you knew that there would be time given to discussion. In short, when you walked into the room, you knew immediately what to expect—and that expectation shaped your behavior. In some cases, you may have turned around and walked out, if that was an option. If it wasn't an option, you may have chosen to sit at the back, or with your friends, or made some other choice. How the room was set up immediately told you what was about to happen, and that began to shape your attitudes and behaviors long before the session facilitator began speaking. How the furniture is organized in the room makes a difference.

In a thinking classroom, how the furniture is organized turns out to make a big difference. Furniture placement sends a message. What is important is that the message that is sent is commensurate with the activity that is intended. That is, a thinking classroom needs to be organized in such a way that says thinking, collaboration, and risk taking are expected. Rows of desks do not achieve this—even if the desks are put together in groups of two or three. Neither do neat rows of tables positioned so students all face the front. These are antithetical to the message we want to send.

> In a thinking classroom, how the furniture is organized turns out to make a big difference.

In my efforts to find the optimal furniture placement for a thinking classroom, I simply showed students pictures of a variety of classrooms and asked them what they thought the teacher and teaching would be like in these rooms. Each of these pictures (see Figure 4.1) showed a classroom with different placement of desks and tables. The first thing to emerge from these data was that straightness equated to order— orderly teacher and orderly teaching. It didn't matter whether the room had desks or tables; if they were arranged along straight lines, then students perceived this to mean that the classroom was organized and well structured. Deeper probing revealed that these assumptions extended to the expectations of the students and their behaviors—that they would also be orderly, that their work and personal spaces would be organized, and that disorder would be frowned upon. While some of the students expressed appreciation for this kind of order, the majority expressed that being a student in such a class would come with a lot of expectations and pressure. In short, straight lines was—good or bad—communicating what was expected in those rooms.

Figure 4.1 Classrooms with different furniture placement.

Sources: Top Left, clockwise recep-bg/iStock.com, skynesher/iStock.com, photo courtesy of Mike Pruner, photo courtesy of Alex Overwijk, photo courtesy of Lisa Poettcker, and photo courtesy of Jamie Mitchell.

Symmetrical furniture placement—placement of desks into a horseshoe or circle, or placement of tables such that all tables are parallel to each other—as it turned out, also conveyed an expectation of order. Because symmetry is often a byproduct of straightness, this result was slower to emerge. The same was true of *fronting*—the placement of chairs so that all students face toward the front of the room. Fronting the room, like straightness and symmetry, communicates that order and compliance are expected in a room. It also communicates that students will be doing a lot of watching and listening. If all three of the furniture placement characteristics—straightness, symmetry, and fronting—are present (see Figure 4.2), then students perceive a room to be very orderly and expect that all activity will be centered on the teacher.

Applying my contrarian experimental methodology, these results told me that we needed to try creating spaces where the arrangement of desks and tables was neither straight nor symmetrical and where chairs were placed in such a way that the room was not fronted. Our investigations, in this regard, consisted of two experiments, the first of which was to show students pictures of a classroom where desks, tables, and chairs were arranged in non-straight, non-symmetrical, and *defronted* fashions (see Figure 4.3). Most students reacted positively to these pictures. Whether students liked or disliked the pictures, however, all students predicted the teacher to be fun and relaxed, and all students thought that there would be a lot of student activity.

Figure 4.2 Orderly classroom.
Source: skynesher/iStock.com

Figure 4.3 Defronted classroom.
Source: Photo courtesy of Alex Overwijk.
Used with permission.

A second result that emerged from this experiment was that we only needed to defront a room in order to also destraighten and desymmetrize it, as long as we defined *defronting* as ensuring that every chair in the room was facing a different compass direction. Doing so automatically ensured that there would be no straight or symmetrical furniture placement.

> We only needed to defront a room in order to also destraighten and desymmetrize it.

Our second experiment was to actually teach in such defronted spaces. Using random groups and vertical non-permanent surfaces added further constraints to the placement of the furniture. First, we needed to arrange the desks and/or tables in such a fashion that groups of three could sit together. Second, we needed to also make sure there was room around the perimeter of the room for groups of three to have easy access to the vertical non-permanent surfaces as students worked, discussed, and moved around the room.

Given that most of the student activity was happening at the vertical surfaces, we were wondering how the furniture placement in the middle of the room would affect student behaviors and attitudes. I didn't have to wait long. The very first room we defronted was a Grade 7 class that had already been doing vertical surfaces for a few weeks. When the students walked in, one of the boys immediately remarked, "Hmm. I guess that's how it is going to be now." When I caught up with him later in the lesson and asked him what he meant by that comment, he told me that the way the desks were placed meant that "the teacher was never going back to teaching the way she used to." The furniture placement, coupled with recent changes in the teacher's practice, communicated that changes were permanent. I will discuss this more in Chapter 15.

> The way the desks were placed meant that "the teacher was never going back to teaching the way she used to."

Even when we defronted classrooms wherein the teacher had made no other changes to their practice, students immediately commented on the desk placement as something indicative of other changes to come. And changes did come. Defronting the room had an immediate effect on both the students and the teacher. Students began to collaborate more, and teachers started talking less. It turns out that how the desks and tables are placed not only sends a message of what is expected, it changes what actually happens. From interviews with teachers, we knew that they placed their furniture to suit their intentions. This, it turns out, is not a one-way street—the placement of furniture also shapes the teacher's intentions. And it shapes their actions.

I already mentioned that we noticed a marked decrease in how much the teacher talked when teaching in a defronted classroom. We also saw a decrease in how much they demonstrated things on the board at the front of the room. Teachers also began to circulate more throughout the room while they were talking and facilitating whole class discussion or answering student questions.

We wondered what other alterations we could make that would have the same power as defronting the classroom. But in fact, further refinements to how desks and tables were placed yielded no better results than the defronted classroom. That was the singularly most effective thing we could do in terms of room organization to induce student thinking.

Figure 4.4 A teacher fields questions from a group in a defronted classroom.

FAQ

Q What do I do about my desk? Where do I put it when there is no front?

A Most teachers have their desk at what has been conventionally called the front of the class. In a defronted classroom, it can go anywhere. However, as part of the defronting process, it should be moved to another part of the room. Otherwise the students will continue to think of where it is placed as the front. Place it somewhere near what used to be the back, but make sure it does not block any vertical surfaces you may wish to use.

Q What do I do about my projector (or interactive white board or clock)? It is currently fixed at what used to be the front of the room.

A There is often not much that can be done about moving these remnants of a fronted classroom. If you can, move it to another wall. Otherwise live with it. The important thing is not to reinforce these as the front. Don't leave a projector (or interactive white board) on unless you are using it. And when it is on, make sure that, as much as possible, you are somewhere else in the room. That is, use your position in the room to contradict the projector's (or interactive white board's) message that it is at the front of the room. If your only remnant of a fronted classroom is the clock, go to a discount store and buy three more clocks.

Q I teach math in a science lab where all the tables are bolted to the floor. How do I defront that room?

A There are three elements that go into fronting a room—how desks or tables are arranged, where the students are seated around these desks or tables, and where the teacher positions him- or herself. Being in a classroom where tables are fixed to the floor defines only one of these three elements. You can still disrupt the other two by positioning chairs on all sides of the tables and making sure that you spend a lot of your time in parts of the room other than what used to be the front.

Q I share my classroom with another teacher who likes the desks to be in rows. How do I defront the room when they do not like that?

A A willing class of students can rearrange the desks in a room in less than a minute. Have your students defront the room as soon as they come in. Just prior to turning the classroom over to the other teacher, have your class refront the room. Some teachers place two small maps of the room on the corner of each desk. The first map shows what the desk arrangement is in a fronted classroom, and the second map shows the arrangement in a defronted class. On each map, one desk in the arrangement is shaded in, telling the student who occupies that desk where it belongs in either configuration. Students come into class, grab a desk, and move it to its correct position.

Q I have big tables in my room. How do I keep students in groups of three when I need more students at each table?

A Sometimes the size of the tables is not commensurate with your desired group sizes. In these cases it is better to have the students squished when sitting than when standing at vertical non-permanent surfaces. So, if you have big tables, put two groups at each. If you have small tables, put one group per table. This will reduce the amount of furniture in your room and give you more space around the perimeter of the room, which will be useful when we get to Chapter 10 on consolidation. Less furniture will also allow for the possibility of an empty space in your room—something that has been shown to be beneficial and will be discussed in Chapter 6.

 **SUMMARY**

MACRO · MOVE

☐ Defront the classroom.

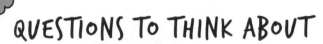

MICRO · MOVES

☐ Cluster desks and tables away from the vertical surfaces.

☐ Position desks and tables so that chairs point in all different directions.

☐ Try not to stand at what used to be the front.

☐ Move around the room when you are talking to the students.

QUESTIONS TO THINK ABOUT

1. What are some of the things in this chapter that immediately feel correct?

2. Think about the way the furniture in your room is currently arranged. Is that for your benefit or the students'?

3. What is it you like about the way the furniture is currently arranged? Why do you like it?

4. Think about other arrangements that you have seen. Why would a teacher prefer that?

5. In this chapter I talked about straightness and symmetry. What else in a classroom, and in teaching practice, might be governed by a desire to have things be straight and symmetrical? What, if anything, do you like about this? What message does this send to students?

6. If we think about the fact that everything we do sends a message to the students, what is the main message that students hear from your practice? Is this the message that you want to be sending them?

7. What are some of the challenges you anticipate you will experience in implementing the strategies suggested in this chapter? What are some of the ways to overcome these?

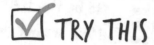 TRY THIS

Grades K–2: Jellybeans

You have 16 jellybeans and four jars.

1. Place the jellybeans in the jars so that each jar has either 3 or 6 jellybeans. Are there some things that are not possible?

2. Place the jellybeans such that each jar has one more than the jar before it. How **many** ways can you do this?

3. Place the jellybeans so that each jar has twice as many as the jar before it. Three times as many.

Grades 3–8: Four Numbers

Select four numbers from 1 to 9 at random. Using these four numbers and any operations, make the values from 1 to 30.

Grades 9–12: Gold Chain

You are backpacking through Europe. You have one month (30 days) left until your flight home, but you have run out of money. However, you have a 50-link gold chain that you bought on your travels, and you have found a hotel that is willing to accept one link per night for payment of room and board. However, the manager wants payment every day, and they are willing to help you out by cutting links for you. The problem is that they want one gold link in payment for every link they cut. How many links will you have left when you fly home?

[Hint: Eventually you want to get the students to think about this problem through the idea of making change. For example, if on a certain day we owe two links of gold, we pay with a length of four links and get two single links as change.]

CHAPTER 5

HOW WE ANSWER QUESTIONS
IN A THINKING CLASSROOM

Much has been written about the effective use of questions from the teacher to direct students' thinking (Andrews & Bandemer, 2018; NCTM, 2014; Smith & Stein, 2018), and I will draw on some of that literature when I discuss the use of hints and extensions in Chapter 9. This chapter is more concerned with how teachers *answer* students' questions and the effect that this has on getting students to think—or not think. By the end of this chapter you will have learned what types of questions students ask and which ones we, as teachers, should be answering.

THE ISSUE

You may be familiar with research showing that teachers ask up to 400 questions a day (Vogler, 2008), but a more interesting question as it relates to thinking classrooms is how many questions teachers are *answering*. And one of the things that became immediately apparent when I was first observing classrooms is how much *time* teachers spend answering students' questions. At first, I saw these answers as background noise to the central focus of a teacher's planning and delivery of a lesson. But as I began to pay closer attention to how teachers answer questions, I realized that this background noise can become overwhelming over time. After an initial study to gauge how answering questions is related to student thinking, I came to the startling conclusion that a typical teacher will *answer* between 200 and 400 questions a day, with some answering as many as 600 questions.

> A typical teacher will answer between 200 and 400 questions a day.

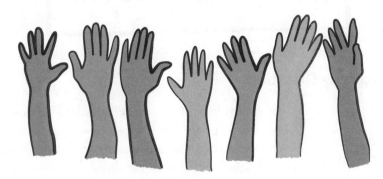

THE PROBLEM

The problem with this is that answering all of these questions is antithetical to the goal of getting students to think. In fact, my research on the other 13 practices was showing that practices that could get students to think were often being undone by teachers answering every question being asked of them. For example, giving students a thinking task is pointless if we then proceed to answer all the students' questions about how to solve it. Likewise, randomizing students into groups of three does not foster collaborative *sense making* when we as teachers give away the *sense made* through our own answers. Yet, answering questions is as much a part of being a teacher as asking questions is a part of being a student. And it's a habit teachers find difficult to change.

Talia   It's so hard for me not to answer every question. I want to help them, and I just want them to like me!

So, what to do? The questions are not going to stop coming. So, what do we do instead of answering them? The answer lies not in whether or not we answer students' questions, but which questions we answer. As it turns out, students only ask three types of questions: *proximity questions, stop-thinking questions,* and *keep-thinking questions.*

```
1. Proximity Questions
2. Stop·Thinking Questions
3. Keep·Thinking Questions
```

Proximity Questions

Proximity questions are the questions students ask when the teacher is close by—in proximity. On the surface, these questions are no different than questions asked in the other two categories. What is different is that the student does not put up their hand or walk across the room to ask it. From the studenting research discussed in the introduction, we learned that in many cases, students ask these proximity questions more for the sake of conforming to the role of student than for the sake of getting an answer. In many instances where I saw proximity questions being asked, immediate follow-up revealed that the information gained from the answer was not being used at all. In fact, in most cases proximity questions consisted of queries about things that students had either already figured out or made decisions or assumptions about. They simply asked the question because it was a habitual studently thing to do when the teacher happened to be standing nearby.

Figure 5.1 A student asks a proximity question as the teacher moves through the classroom.
Source: Wavebreakmedia/iStock.com

From role theory (Horowitz, 1967), we know that the drive to conform to socially defined roles can be very strong, even if totally subconscious. In a classroom, the dominant roles are that of student and that of teacher. Asking a question, one of the most studently things that a student can do, cements their role as student in the eyes of their teacher. Conversely, answering a question is one of the most teacherly things a teacher can do, thereby solidifying *their* role. The drive to conform to these roles is most obvious in moments when a teacher catches a student doing something they should not be doing. See if this sounds familiar:

> It's toward the end of the lesson, and you have given your students some seat work to complete before they leave. You notice that one of the girls near the back of the room is paying an awful lot of attention to something in her hands. You suspect she is doing something on her phone, but you can't quite see what

it is, as her textbook is blocking your view. So, you walk up the aisle toward her to see what is going on. The girl notices your approach and makes a few brief movements with her hands. As you get closer to her you can finally see over her textbook and you notice her phone, set to a calculator app, lying on her desk. Now that you are there, she asks you, "So, for Question 11, are we supposed to find all the answers, or just one answer?" You respond to her that she is supposed to find all the answers and walk away satisfied that not only is she on task, she is ahead of most of the students in the class.

From my position at the back of the classroom, however, I can see that your first instinct was correct—she was on her phone. And not on the calculator app that you saw, but communicating with her friends through a social media application. As she saw you approach, she quickly switched over to the calculator app—a perfect subterfuge. She has completely managed to cover her tracks, and she knows, as do you, that you would be hard pressed to prove otherwise. So why the question? She had gotten away clean.

She asked the question because, whether you knew it or not, she knew she was out of position—acting outside of her role. The question, the answer to which she did not need, was the quickest way to get back into her socially defined role of student. By answering it, you not only reaffirmed your role as teacher, but also acknowledged that she was in her role as student.

This is not to say that there were not instances of students asking genuine questions when the teacher was close by. This did happen. And when it did, follow-up with the teacher often revealed that the students who asked the proximity questions were either shy or had very good work habits. In the former case, these were students who did not want to draw attention to themselves by raising their hand or getting out of their seats to approach the teacher. Instead, they waited until the teacher was close by and then, very quietly, asked their questions. Teachers seemed to have a sense of this and were drawn to these students in anticipation of their questions, either from prior experience or very subtle signals from these students. In the latter case, rather than take the time to put up their hand and wait for the teacher to come to them, or walk across the room and wait in a line of students with questions, these students chose to save their questions for when the teacher was close by. In the meantime, they proceeded with the rest of their work. Regardless, the number of students with legitimate proximity questions paled in comparison to the number of proximity questions asked for the purpose of establishing, or reestablishing, roles.

Stop-Thinking Questions

The second type of question that students ask is called a *stop-thinking question*. These questions can take the form of "Do we have to learn this?" or "Is this going to be on the test?" More often, however, they take varying forms of "Is this right?" The question could be about an answer they have come up with, their progress on a thinking task, or the way in which they are following your instructions. Regardless, these questions are motivated by the reality that, for students, thinking is difficult, and it's hard to decide for themselves that what they are doing is correct. If they can just get you to do that for them, their life would be so much easier. So students ask this question with the hope that you will answer it, and they can stop thinking.

Keep-Thinking Questions

Keep-thinking questions, on the other hand, are asked by students so they can continue to engage with the task at hand. These are often clarification questions or questions about extensions the students want to pursue. Students who ask these questions are motivated to keep going— keep working, keep thinking.

> Can we get the next one [question]?
> When you say numbers that add to 25, do we have to stick to whole numbers?
> We want to try to solve this for the negative case as well. Is that ok?

When we get to Chapter 9, we will see that students will eventually stop asking you these types of questions as they gain confidence in their thinking and start creating extensions for themselves.

TOWARD A THINKING CLASSROOM

Answering these proximity or stop-thinking questions is antithetical to the building of a thinking classroom.

It turns out that of the 200–400 questions teachers answer in a day, 90% are some combination of stop-thinking and proximity questions. How many of these are proximity questions depends on how much the teacher circulates around the room during a lesson. Teachers who spend a lot of time circulating receive many more proximity questions than those who circulate very little or not at all. Regardless, answering these proximity or stop-thinking questions is antithetical to the building of a thinking classroom. In the best case, the answer would be redundant. In the worst case it would shut down thinking.

The only questions that should be answered in a thinking classroom are the small percentage (10%) that are keep-thinking questions. But this raises two new problems—how to quickly discern the types of questions being asked and how *not* to answer 90% of them?

The first of these issues turned out not to be a problem at all. The teachers I was working with quickly learned to discern a keep-thinking question from other types of questions. One of the things that helped with this was the realization that almost all questions asked in the first few minutes of a thinking task are either proximity questions or stop-thinking questions—neither of which needs to be answered. Although the questions being asked early on in a task usually appear as clarification questions, in truth they are often being asked to avoid having to do the hard work of discerning what is being asked and making decisions about things that are perceived to be ambiguous.

Once the task is up and running, it is equally easy to discern the nature of the questions being asked if you just keep in mind what the downstream effect of you answering the question will be. Are they asking for more activity or less, more work or less, more thinking or less? But, watch out. In an environment in which you are only answering certain types of questions, students get very inventive in how they formulate their questions, often forming statements with an accompanying [implied question] and lifting their eyebrows.

We are thinking this is correct! [What do you think?]
This is correct! [Right?]
I think we are going the right way! [Right?]

Don't be fooled by these pseudostatements. If their tone is inviting a response from you, they are really asking a question—and by the nature of the disguise, it is almost always a stop-thinking question. Focusing on the consequences of responding to their statement quickly helps you discern the intention behind the question. Are they trying to get you to help them to stop or keep thinking? If in doubt, assume it is to get you to help them stop thinking.

Figure 5.2 Elementary students ask questions as teachers approach their work stations.
Source: Photo courtesy of Sheri Stashick. Used with permission.

What turned out to be more difficult was figuring out what to do in place of answering a proximity or stop-thinking question. Students can be very persistent in their efforts to get you to help them reduce their workload, and how you respond to this is important.

> Students can be very persistent in their efforts to get you to help them reduce their workload, and how you respond to this is important.

Working with a team of eight teachers, we came up with a list of 10 things to say in response to a proximity or stop-thinking question.

1. Isn't that interesting?

2. Can you find something else?

3. Can you show me how you did that?

4. Is that always true?

5. Why do you think that is?

6. Are you sure?

7. Does that make sense?

8. Why don't you try something else?

9. Why don't you try another one?

10. Are you asking me or telling me?

Each of these suggested responses is a variation of answering a question with a question. Some of the teachers I was working with became quite proficient at using this list, and for them, these responses were effective at redirecting proximity and stop-thinking questions. But these teachers were among the minority. For the majority of teachers, answering a question with a question became a slippery slope into revealing more than what they initially intended. Consider the following interaction.

Teacher	Why don't you try something else?
Students	Like what?
Teacher	Maybe you need to consider the cases where x is negative.
Student	You mean like this?
Teacher	Right!

What we found was that answering a question with a question (and only a question) was only effective when it was immediately followed by the teacher walking away from the students, with no other statements or suggestions being made. In fact, this was so clear that we decided to try this strategy on its own. We would just walk away. This turned out to be infuriating to students and did cause some negative backlash. But after two weeks we also noticed it caused a sharp decrease in the number of proximity or stop-thinking questions being asked by students—in some cases reducing the number to fewer than 30 questions a day. As students began to realize that their questions weren't going to be answered, they stopped asking them . . . except in the primary grades.

If a six-year-old asks a question and it is not answered, they ask it again. If it is still not answered, they ask it again. And if it is still not answered they do something that a 16-year-old does not. They reach out and touch the teacher—tap them on the arm or pull on their clothing. And if the teacher walks away, they follow. I have multiple videos of kindergarten and Grade 1 teachers walking around the room with a row of little ducklings following them. As soon as the teacher stops walking, they are immediately surrounded by these little ducklings tapping and tugging at them.

These hilarious episodes of primary teachers trying to not answer students' questions by walking away prompted us to explore some nuances of this strategy at all grades. For students there is a big difference between having their question heard and not answered, and having their question not heard. The primary students were reacting to the latter of these. They assumed that their questions had not been heard. No one likes to be ignored. So, we made a modification to the walking away strategy. Instead of walking away when a proximity or stop-thinking question is being asked, we would instead look at the student and smile as they asked their question. Then we would walk away.

> For students there is a big difference between having their question heard and not answered, and having their question not heard.

This turned out to have a huge effect on the perceptions of students at all grades. Instead of feeling ignored, they now knew that they had been heard and that the teacher's decision to not answer them was deliberate. Many students took this to mean that they needed to do more work. Over time, the students began to see the smile and walk away as a sign that the teacher had confidence in their ability to resolve the question on their own. There were still a few students who were frustrated by these encounters. But they were thinking more—or no longer having the teacher do their thinking for them.

When coupled with the aforementioned building thinking classroom practices, students perceive the smile and walk away as you having confidence in their group, and the room as a whole, to resolve their question. This is not to say that students shift their questions from the teacher to their groupmates. Students do not ask proximity questions of their peers—there is no need to establish their role as student within the group. They do still ask their groupmates if "this is right," however. But because their groupmates do not have the same authority as a teacher, these questions are asked and answered with a level of tentativeness that keeps the thinking going.

 What do we do when a student, or a group of students, insists that I answer a stop-thinking question?

A This happens most often in the context of an *Is this right?* question. The easiest way to deal with this is to call it like it is— "I'm not going to answer that question". Then tell them why— "Me telling you that it is right is worth almost nothing. If you can tell me that it is right, however, that is worth everything." And, then tell them that you have confidence in them—"And I believe that you will be able to tell me if this is the right answer. So, keep going." In some instances, you may wish to couple this with a hint (see Chapter 9).

Q How do I tell the difference between a keep-thinking question that is asked as I am moving round the room and a proximity question?

A The key difference between these types of questions has to do with the activity of the student, or group of students, at the time the question is asked. If the students are busy working away at whatever task is at hand, any questions they ask tend to be proximity questions—"For Question 3, were we supposed to find all the answers?" or "Are we doing this right?" Keep-thinking questions, on the other hand, tend to come when students are at an impasse and need something from you to move forward. In some cases, this is a request for a hint (see Chapter 9)—"We're having trouble here. Were we supposed to do this for all the possible sizes?"—or an extension— "Are we supposed to now look at the general case?"

Q Should I tell my class about the three types of questions they ask and that I am no longer going to answer proximity and stop-thinking questions?

A As it turns out, this is the first practice where we experimented with talking to students about what we were doing. When done correctly, two interesting things happen. The first is that students started to self-regulate the types of questions they were asking. In this regard, we saw a huge decline in proximity and stop-thinking question, coupled with a small uptick in the number of keep-thinking questions being asked. The second thing that happened was that students started using the language of the three types of questions to moderate their peers—"Dude! She's not going to answer that. That's a stop-thinking question."

The challenge was doing it correctly. In this regard, we learned that talking to students about the practice before initiating it almost always resulted in challenges when the implementation began. From interviews with students, we learned that students perceive pretalks to mean that the teacher is asking them to behave themselves during the implementation. By extension, this meant that the teacher was asking their permission to make this change, and in the minds of some students, this gives them the power to decide whether this is something they want or not. On the other hand, talking to students about the practice after two weeks of not answering proximity and stop-thinking questions, coupled with a lot of smiling and walking away, was met with very positive reactions—"So that's what is going on!" This was not perceived as asking permission, but rather an explanation for something that had already become an implicit part of their classroom norms. Students also appreciated the peek behind the curtain of teaching—"It's cool that he told us that! It's like he is really thinking about what he is doing."

This distinction between pre- and post-implementation discussion played out the same every time we experimented with talking to the students about why we were doing what we were doing—irrespective of the practice we were talking about. Although many of the same things were said in both instances, the students perceived preimplementation conversation as asking permission and postimplementation discussing as inviting them into the reasons behind the practice.

> Students perceived preimplementation conversation as asking permission and postimplementation discussing as inviting them into the reasons behind the practice.

Q I can see how smiling and walking away, although infuriating for students at first, can become a good way to avoid accidentally answering a proximity or stop-thinking question, and thereby a good way to get students to keep thinking. But does it work for all students?

A Yes and no. It works in that you are not letting them stop thinking. It doesn't work for all students in the sense that there are students who cannot get past the fact that you have not answered their question. This may be because they are insecure about their own abilities, have learned helplessness, or have a spectrum disorder—such as obsessive compulsive disorder—that does not allow them to move forward without resolution. Alternatively, you may have students who become incensed at your deliberate disregard for their question. After all, they have a lifetime of experiences with teachers answering their questions. Or, it may be a combination of these factors. Regardless, you need to read the situation and know when a nod, a wink, or an encouraging remark—"I have complete confidence that you can figure this out"—is needed. Keep in mind, however, that when you open your mouth, you may be overcome with an almost undeniable desire to answer their question. It is in our nature as teachers.

> You need to read the situation and know when a nod, a wink, or an encouraging remark is needed.

Although we had great success with smiling and walking away in K–2 classrooms, there were some settings where several students didn't know how to process the signals that these actions were meant to send. In these instances, teachers would smile, say a few encouraging words, and then walk away. Over time, they could stop with the encouraging words and just smile and walk away.

What is important, is that you do not answer proximity or stop-thinking questions *and* that you read the situation so as to be able to give the best response when such questions are asked.

Q How do parents react to the teacher not answering questions?

A The answer to this depends on who tells them. If you leave it to the students to tell their parents that you are not answering their questions, I think you can imagine what their reaction will be. Students, at the best of times, are not great at communicating the nuances of what happens in the classroom. Couple this with a parent filtering what they hear from their child through their own experiences as a student, and you are likely to get an e-mail or a phone call. If, however, parents hear from you that you are doing everything to encourage student thinking and that you will be supporting that thinking through how you selectively answer and don't answer student questions, the response tends to be much more amicable.

Q I have been implementing thinking classrooms for a while, and there is no way I am getting 200–400 questions a day. What kinds of classrooms were these numbers coming from?

A The 200–400 questions a day data comes from classrooms where none of the thinking classroom practices are being implemented. As you begin to implement these practices, these numbers dramatically decrease, in some cases going down to close to zero questions asked in a single period.

Q So, smiling and walking away is the strategy *after* the question has been asked. Are there any strategies that can be used *before* the question is asked to prevent it being asked at all?

A There are three strategies that we have played with—the first of which involves strategically reducing proximity. The first three to four minutes after the first task is given is, by far, the period of time in a thinking classroom when the greatest number of questions is asked. During these three to four minutes, stay in the very center of the room, as far away from the students as possible. By distancing yourself from the students, you reduce proximity questions.

The second method is to not answer *any* questions asked by an individual student. Individual students ask their group members questions; groups ask the teacher questions. So, if a student approaches you with a question, respond with, "What did your group members give as an answer?" or "Did you ask your group?" If the answer is that the group doesn't know, then follow the student back to the group to hear exactly what the question is. This is not to say that you will answer their question. You might give a hint, or you might just smile and walk away.

The third method is to lead with your own questions when you approach a group. "What are you doing here [pointing]?", "Can someone explain what is happening here?", "What do you know so far?" By leading with a question you are controlling the conversation, and it means that if you give encouragement or a hint or a smile, you are *not* doing so in reaction to one of their questions.

SUMMARY

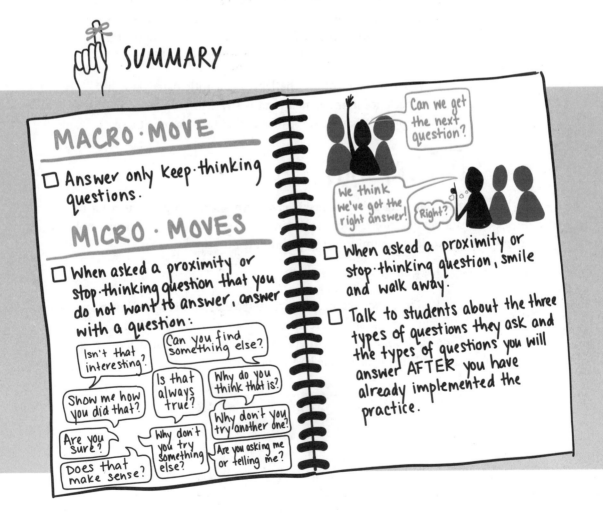

QUESTIONS TO THINK ABOUT

1. What are some of the things in this chapter that immediately feel correct?

2. The introduction talked about institutional norms being a potential source of student disengagement and lack of thinking in the classroom. This chapter talks about the way we, as teachers, answer questions as contributing to students' not thinking. In what other ways do our interactions with students reduce or remove their need to think?

3. Many of the practices for building thinking classrooms discussed to this point are ways in which we can create environments that get students to think. The practice discussed in this chapter, in many ways, is the opposite of this. In this chapter, you learned about ways to avoid doing things that stop thinking. What other practices stop thinking?

4. What are some of the challenges you anticipate you will experience in implementing the strategies suggested in this chapter? What are some of the ways to overcome these?

✓ TRY THIS

The following thinking tasks have been shown to generate a lot of student questions. Therefore, using these is a great way to practice not answering proximity and stop-thinking questions.

Grades K–4: Ice Cream Cones

The Ice Dream ice cream shop has 10 flavors of ice cream. How many different two-scoop ice cream cones can you make? What if there were 11 flavors? What if there were 12 flavors? What if it were 20 flavors? What if each cone had at most three scoops?

Grades 5–8: Palindromes

A palindrome is something that is the same forward as backwards—like *mom*, *dad*, *race car*, *I prefer pi*, et cetera. Numbers can also be palindromes—like 141, 88, 1221, et cetera. Now, consider the number 75. 75 is not a palindrome. So, reverse it and add it to itself: 75 + 57 = 132. 132 is also not a palindrome, so do it again: 132 + 231 = 363. 363 is a palindrome. So, we stop, and we say that 75 is a depth-2 palindrome (because I had to do the process twice to get to a palindrome). Find the palindrome depth of all two-digit numbers.

Grades 9–12: Wine Chest

Mr. Snooty loves red wine. So much so that he drinks one bottle of wine a day. But he is very particular about his wine. First, it has to be the right type of wine. Second, it has to be the right temperature. And third, it cannot have been exposed to light more than five times. To make sure it is the right type of wine, Mr. Snooty goes to his favorite wine store, which is very far from home. To make sure that the wine is at the right temperature and not exposed to light, Mr. Snooty built two temperature-controlled wine chests in his house—one much bigger than the other. How often does Mr. Snooty have to go to his favorite wine store? Mr. Patooty shops at the same store as Mr. Snooty, likes his wine at a certain temperature, but will not drink wine that has been exposed to light more than ten times. How often does Mr. Patooty have to go to his favorite wine store?

CHAPTER 6

WHEN, WHERE, AND HOW TASKS ARE GIVEN IN A THINKING CLASSROOM

In Chapter 1 you learned about the qualities of tasks for thinking classrooms and how these qualities are important to initiating and maintaining thinking. Having such tasks, although necessary for the building of thinking classrooms, is not enough. What you do with them is far more important. In Chapters 2 and 3 you learned about the research that showed that these tasks come to life when students work on them in random groups on vertical non-permanent surfaces. In this chapter we will look at the research results that show that the more subtle practices of when, where, and how the tasks are given is as important as the quality of the task itself.

> The more subtle practices of when, where, and how the tasks are given is as important as the quality of the task itself.

THE ISSUE

The internet is full of resources of rich tasks. Whether you look through the archives of NRICH (nrich.maths.org), NCTM Illuminations (illuminations.nctm.org), or simply type *problem of the day* into your favorite search engine, you will find endless lists of potentially good thinking tasks. Yet, one of the most frequently asked questions I get is still, "Where can I get good tasks?" Where is the disconnect? With the abundance of good tasks, why would people ask me this question?

My suspicion is that this question is actually a proxy for deeper, more imperative, questions like "Where can I get good tasks that help me teach the curriculum?" or "Where can I get good tasks that engage my students?" The answer to the first question was briefly discussed in Chapter 1 and will continue to be elaborated on in Chapter 9. The answer to the second question is fundamentally what this entire book is about—and the answer lies not in the task, but what we do with it.

THE PROBLEM

Consider the palindrome task given at the end of Chapter 5. By all accounts that is a great task for getting students to think. We have used it in hundreds of Grade 5–12 classrooms, and, almost without fail, it generates extended periods of deep mathematical thinking. And now you too have this amazingly rich thinking task. The question is, what are you going to do with it? Assuming that you want to use it with your students, how are you going to give it to them?

In all the research we have done, out of all of the hundreds of microexperiments we ran, and from the hundreds of interviews we conducted, this one question was the thing that teachers thought the least about. How are you going to give the task? As it turns out, teachers tend to give tasks in one of three ways—they project it or write it on a vertical surface, they give it as a handout, or they assign it from a textbook or workbook. Of these, which is the worst? Which method, when the exact same task is used, generates less thinking than any other way? Even though this is the part of their practice that teachers think the least about, you likely know the answer. We all know the answer—the textbook/workbook generates less thinking than the other two methods.

> Students have been socialized to believe that questions are assigned from the textbook or workbook after they have first been shown how to do them.

Students, especially students above Grade 7, have been socialized to believe that questions are assigned from the textbook or workbook after they have first been shown how to do them. This is not an unreasonable assumption on their part, as questions are typically assigned from books near the end of the lesson—after a lesson full of worked examples. This assumption, however, interferes with the way they engage with the task. Rather than approaching questions in the book as something to think about, they approach them as something to be answered by mimicking the examples from the lesson and their notes. And when this does not work, rather than think about the question, they put up their hand and ask for help on how to do it. This same baggage does not accompany a task that is projected or written on the board or given as a handout—even if it is the exact same task.

How we give the tasks matters, and it turns out to matter a great deal. The same is true of *when* we give the task, and even *where* in the room we give the task. And like the textbook/workbook example, how we naturally do it, how we have been taught to do it, often produced the lowest levels of thinking in our research.

TOWARD A THINKING CLASSROOM

Before we get to discussing how to give a task, you will first see the results of when and where to give the task.

When to Give the Task

One of the earliest results from the research was that the same task given either in the middle of the lesson or near the end of the lesson produced much worse results than if it was given right at the beginning of the lesson. There are two main reasons for this—the first of which has to do with the students. As discussed in the previous chapter, students prefer to occupy lower energy states. So, if a lesson begins with the low-energy state of passively receiving knowledge in the form of a lecture or taking notes, it is much harder to then raise their energy level and get them to start thinking. This was obvious in the data through students' complaints, questions, and slow starts whenever we gave them thinking tasks in the middle or near the end of the lesson. A class given these same tasks at the beginning of the lesson came to it with energy, enthusiasm, determination, and a greater sense of self-reliance. Even when we ran tasks on the same students at different points of the lesson, we saw a huge difference between a task given early versus late in a lesson. It wasn't about the task, and it wasn't about the students. It was the timing of the task that made a difference.

> If a lesson begins with the low-energy state of passively receiving knowledge in the form of a lecture or taking notes, it is much harder to then raise their energy level and get them to start thinking.

The second reason for the shift in engagement from the beginning of the lesson to the middle or end of the lesson was about the teachers. From our data it became clear that the longer the lesson progressed before a thinking task was given, the more likely the teacher would begin to preteach the task in some way. Sometimes this was explicit with parallel worked examples. Sometimes it was more subtle with emphasis on useful representation, organization, and strategies. Interviews with teachers after such events were mirthful as they laughed at themselves.

| Samantha | Ha. I just can't help it, I guess. |
| Stanley | Humph. I guess that's how we're wired. |

> The idea of preparing our students for what is to come is so engrained in the fabric of teaching that, even when we know it is counterproductive to thinking, it is difficult to stop.

The idea of preparing our students for what is to come is so engrained in the fabric of teaching that, even when we know it is counterproductive to thinking, it is difficult to stop.

This preteaching, coupled with the initial passive positioning of the students, undermines the effectiveness of a task to generate thinking in an almost linear fashion (see Figure 6.1). The further into the lesson the teacher waited before giving the task, the less effective it became. Pushing this line of research further, we determined that the teacher has three to five minutes from the beginning of the lesson to give the task before this deterioration begins. Interestingly, what defines the beginning of the lesson, in this regard, is not when the bell rings, but when the teacher begins to address the class as a whole.

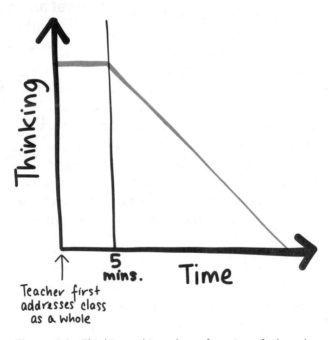

Figure 6.1 Thinking achieved as a function of when the task is given.

You may be thinking that three to five minutes is doable for review lessons. But when a lesson begins with the introduction of new content—content that is needed to do the first task—I would wager you're thinking this is too little time. I can assure you that it is not, but you are going to have to wait until Chapter 9 to find out why not. In the meantime, just know that the longer you talk, and the longer they listen, the less likely you are going to be able to get them to think.

Where to Give the Task

Student passivity is accentuated not only by how long we talk before giving them a task. Where the students are located while we are talking to them also has an impact on the degree to which they are put into an active or passive state. For example, having students sitting in the desks, while the teacher talks, creates a low-energy and passive environment for students. In comparison, having students stand, loosely clustered around the teacher, creates a higher-energy and active environment for the students. This was confirmed in the data. If students were left in the desks in the lead-up to being given a thinking task, they were much slower to start and had more proximity and stop-thinking questions. In contrast, when the students stood and gathered around the teacher, they were quicker to get going on the task and were less likely to ask questions.

Figure 6.2 A teacher gives his students the task among standing students.
Source: Photo courtesy of Judy Larsen. Used with permission.

> Sitting in a desk is powerfully associated with direct instruction, passive learning, and non-thinking behavior.

From a physiological perspective, this makes sense. Standing versus sitting requires a more active use of core muscles and increases blood flow. From a psychological perspective, sitting in a desk is powerfully associated with direct instruction, passive learning, and non-thinking behavior. This association is not only embedded within the institutional norms of school but is also part of the lived experience of students as young as Grade 2 or 3. This was confirmed in interviews wherein we showed students pictures of students listening to the teacher while sitting in desks, sitting on the floor, and standing around the teacher. When asked what they thought class was like for each picture, almost every student indicated a more positive affective response to the pictures where students were not sitting in desks. Even a picture of some students sitting on lab tables as they listened solicited a more positive response than the picture of students sitting in desks. Deeper probing revealed that students' positive attitudes were associated with their sense of engaging teaching.

In a more simplistic study, we documented how many high school students were looking at their cell phones while sitting versus standing while listening to the teacher. The results were remarkable. In the case of sitting, upwards of 50% of students looked at their cell phones at least once in a five-minute interval. For the students standing, that number dropped to less than 10%. This likely has something to do with the sense of anonymity discussed in Chapter 3, but it also has to do with the degree to which students are engaged. Whereas disengaged students look for distraction, engaged students are not distracted.

Taking these influences together, it was clear that the three to five minutes that a teacher has before sending students off to do the first thinking task are much better spent talking to the students while they are standing in loose formation around them.

How to Give the Task

> Nothing came close to being as effective as giving the task verbally.

As mentioned, giving the task through the medium of the textbook or workbook came with a lot of baggage and produced the least amount of thinking and greatest amount of proximity and stop-thinking questions. Incidentally, giving the task in the form of a worksheet also produced very poor results, as it was associated much more with *getting done* than with thinking. Of the three most common ways teachers give a task, that left only projecting or writing the task on a vertical surface. In our experiments with the different mediums for giving tasks, this was the best option—or rather, it was the least bad option. At least writing on the board was something that worked well with having students standing in loose formation around the teacher. However, writing on a vertical surface was far less effective than what turned out to be best—giving the task verbally.

This was a shocking result. Not only does it go against decades of norms, it also goes against teachers' instincts. However, before we get into the details of the results, some clarification is needed as to what is meant by *verbal* in this context. Whereas the essence of the task is given verbally, the details of the task—quantities, measurements, geometric shapes, data, long algebraic

> Verbal instructions are not about reading out a task verbatim.

expressions, et cetera—are written on the board *as the teacher speaks*. Verbal instructions are not meant to be about the students having to remember details, it is about having them hear the nature of the question. Likewise, verbal instructions are not about reading out a task verbatim. Rather, they are about unwinding the task through narrative, discussion, dialogue, and potentially working through a model of what is being asked with the students. So, for example, rather than just reading out the palindrome task as written in the end of Chapter 5, the teacher would gather the students near a vertical non-permanent surface and go through something resembling the following script.

Teacher	Does anyone know what a palindrome is?
Class	A word that is the same forward as backwards.
Teacher	Can you give me an example?
Class	Mom, dad, Hannah, race car.
Teacher	Ok. Good! Can someone give me a number that is a palindrome?
Class	1221, 25452, …
Teacher	What about 99? Or 8? Ok. Someone give me a two-digit number that is not a palindrome.
Class	14.
Teacher	Ok. 14 is not a palindrome. So, what I am going to do is take 14 and add it to its reverse. 14 + 41

	[teacher writing on the board], which is 55. Is 55 a palindrome?
Class	Yes.
Teacher	Yes, it is. So, I stop. What if I start with 48?
Class	48 + 84 = 132 [teacher writing on board].
Teacher	Is 132 a palindrome?
Class	No it's not.
Teacher	No it isn't. So, I do it again. 132 + 231 = 363 [teacher writing on the board]. 363 is a palindrome so I stop.
Teacher	So, when we started with 14, we needed to reverse and add once before we got a palindrome. Therefore, 14 is called a depth-1 palindrome. For 48, we had to do it twice before we got a palindrome. Therefore, 48 is called a depth-2 palindrome.
Teacher	Your task is to come up with the depth of all two-digit numbers.
	[Teacher forms random groups.]

Of course, the script may not unfold exactly this way, but with some prodding and pulling you will be able to recreate a close approximation. There are three things to notice in this dialogue.

1. The task is not given until the groundwork has been presented.

2. The groundwork in no way reduces the thinking that the students will have to do.

3. If a student walks into class late and looks at the board, they will have no clue what the task is.

What is on the board is meaningless without the accompanying verbal dialogue and instructions. This turned out to be the definition of what separated verbally given tasks from textually given tasks—the textual residue of giving the task is not enough to discern what the task is. Yet, the textual residue is important. Not only does it reduce the cognitive load of trying to remember details, but it also provides visual cues to go with the verbal instructions and to refer back to as an anchor.

I imagine that you have loads of what-if and yeah-but questions right now. We'll get to those at the end of the chapter. But first let's look at some of the research results of student thinking when tasks were given verbally. One of the tasks I used in this part of the research is the tax collector task:

> I have 12 envelopes, numbered 1 to 12. Each contains a number of dollars equivalent to the number on it. The game starts with you taking one of the envelopes—the money inside of which is yours to keep. The tax collector will then take all of the remaining envelopes whose number is a factor of the envelope you took. The tax collector must be able to take at least one envelope every turn. Play continues until you can no longer take an envelope, at which point the tax collector will take any remaining envelopes. What is the most amount of money that you can get?
>
> Source: Adapted from Diane Resek task "The Tax Collector" (2007).

This is an amazingly rich task that can be used with students as young as Grade 4. In the research, we compared student thinking on this task given textually, as above, with their thinking when the task was given verbally, with students standing around their teacher, according to the following script.

Teacher	So, these are 12 envelopes, each one with some money in it. [Teacher draws 12 rectangles on the board and writes $1, $2, $3, etc. inside each rectangle.] This is your money. I am just holding it for you. But, you can have any of these envelopes whenever you want. You just have to ask for it. So, which one do you want first?

Students	The $12.
Teacher	OK. There you go. [Teacher pretends to hand an envelope to a student and then circles the $12 envelope on the board.] This envelope is now gone.
Teacher	Right [snapping their fingers]. I forgot to tell you that we have to pay taxes on this money. And because you took the $12, the tax collector will take the $1, the $2, the $3, the $4, and the $6 envelopes [teacher crossing out the envelopes as they say these numbers]. Why does the tax collector take these envelopes?

Students   They are the factors of 12 [or the numbers that divide into 12 or go into 12].

Teacher   Right. The tax collector takes the factors of whatever envelope you took. Ok, so which envelope do you want next?

Students   [Laughing] The $11.

Teacher   OK [snapping their fingers]. But I forgot to mention that the tax collector always wants some taxes. So, when you choose an envelope there must always be at least one envelope for the tax collector to take—one factor for the tax collector to take. So, can you take the $11?

Students   No. there are no factors of $11 left.

Teacher   OK. So, which envelope can you take?

Students   The $10.

Teacher   OK. You take the $10. [Teacher pretends to hand an envelope to a student and then circles the $10 envelope on the board.] What does the tax collector take?

Students   The $5 [teacher crosses out the $5].

Teacher   OK. What next?

Students   Nothing. There is nothing else we can take.

Teacher   OK. Now, the tax collector is very kind and does not want to see anything go to waste, so they will take the rest of the envelopes. [Teacher crosses out the $7, $8, $9, and $11.]

Teacher   So, you got $22 in total. That is NOT good. Your job is to do better than $22.

[Teacher forms random groups.]

The narrative and verbal form of the tax collector task contains all of the elements of the text of the original task, but it also adheres to the three characteristics previously highlighted—the task is not given until the groundwork is established, the laying of the groundwork does not diminish the thinking required, and what is written on the board is incomprehensible to anyone who has not heard the verbal component of the task. This particular task also highlights a fourth characteristic that is present in some tasks—the constraints of the task emerge out of, and after, actions have been taken. So, the idea that the tax collector takes the factors is revealed only after the students have selected their first envelope. The idea that the tax collector must always get at least one envelope is revealed after the students try to take an envelope without factors.

These two very different presentations of the tax collector task—textual and verbal—gave us a context in which to compare the differences in students' behaviors when they work on a task that is given textually versus verbally. One of the big differences we observed was the time it takes groups to get to the mathematics inherent in the task. For example, when tax collector is given verbally as above, we  hear groups talking about starting with a prime number in the first 60 seconds of them working on the task. The same utterances are not heard for 10–12 minutes when the task is given textually. This is a huge difference.

When we analyzed what was happening in those first 10–12 minutes, some common patterns emerged:

1. The students spent a fair bit of time silently reading the task—or pretending to read the task. When they finally began to discuss the task with each other, they started by talking about the words—"What does it mean by the factors?"

2. They talked about constraints—"Does the tax collector take all the factors or just one?"

3. They tried to solve the task while simultaneously rereading and renegotiating what the rules presented in the text were.

Each of these things was seen to be very challenging, and many of the students quit. Those who did not quit then started talking about the mathematics in the task, and that is when we observed them, finally, starting to talk about prime numbers, squares of prime numbers, the number of factors, et cetera. On the other hand, when we observed students who received the task verbally, they *began* by talking about the mathematics. Somehow, the posing of the task verbally cut through all of the words and positioned the task in the minds of the students in such a way that they could immediately start thinking about the mathematics.

It would be natural to assume that this is because the narrative dispenses with the reading, negotiating the meaning of the words, and the discussion of the meaning of the constraints. And this is true, of course. But what is interesting is that the same differences were observed in tasks that require no modeling of the task on the board.

Another major difference between how students work on textual versus verbal tasks is the number of questions asked of the teacher. When the task is given textually, students ask lots of questions, primarily during the phases where they are reading, discussing words, and discussing constraints. If I go back to the story of Jane in the introduction, this was one of the first observations I made— Jane was running from student to student answering their questions. The degree to which these questions were answered, and how quickly they were answered, had a lot to do with whether or not the students persisted, or not, through the aforementioned three phases. In the data, I saw cases where students asked so many questions that the teacher called the whole class to attention so that they could go over what the task is asking—*verbally*. That is, when the textual format was not working, several teachers were driven to shift to verbal means to help the students get to the mathematics.

> Giving tasks verbally produced more thinking—sooner and deeper—and generated fewer questions at every grade level, in every context, and even in classes with high populations of English language learners.

In summary, the results showed that giving tasks verbally produced more thinking—sooner and deeper—and generated fewer questions at every grade level, in every context, and even in classes with high populations of English language learners. That is, there was no context in which giving a task verbally led students to perform worse than giving it textually—whether on a board, on a worksheet, or in a textbook/workbook.

FAQ

Q I don't have a large open space in my classroom where all the students can stand while I give the task. What do I do?

A The research showed that such a space was not necessary. The students do not all have to stand between you and the desks. There can be students standing behind desks as well. The important thing is to have them standing and to have them clustered in one area. It also helps if you vary where this cluster will be from day-to-day. You will see in Chapter 10 how informal these clusters can be when you are debriefing an activity.

Q If the students are standing, how do they write down some of the details that I am putting on the board—quantities, measurements, geometric shapes, data, long algebraic expressions, et cetera—that they will need to work on the first task?

A They will not need to do this if you write these things up on a board. So, write these details up high on the board where everyone can see them, both when the students are in their loose formation around you and when they get to the vertical non-permanent surfaces they will work on with their groups. Then draw a box around the details, and ask whatever group ends up there to not erase them. Some teachers have a dedicated board in the room that only they write on (not in the "front" of the room), and some teachers use static cling whiteboard sheets for this purpose and then move them to a wall where no group is stationed.

Q I have some students who will not be able to absorb the question if I give it verbally. Should I give it to them textually, or should I also project it textually while giving it verbally?

A Our research clearly showed that giving a task verbally and textually at the same time produces the same results as giving the task textually only. With the presence of text, the students will be drawn to the written words first and not listen to your verbal unwrapping of the task. With respect to the few individuals in your class who you worry will not be able to function verbally, odds are that they are not great at decoding text either—very few students are. All students are verbal learners long before they are textual learners.

This does not go away when they learn to read. What those students you are thinking about may not be good at is taking verbal instruction in large-group settings. One-on-one, they are likely fine as verbal learners. When they get to their small groups, there will be lots of opportunities for their group members to reexplain the task in a more focused setting. As a teacher, you know for which students such reexplaining may be necessary, and you can observe to make sure this is happening.

Q What if most of the students don't understand what the question is?

A The research showed that when the boundaries are very porous (Chapter 2) and there is a lot of autonomy in the room (Chapter 8), then only about 20% of the students need to understand the task. Knowledge mobility takes care of the rest.

Q What do I do if I see a group has completely misunderstood the question?

A As a teacher, you will be required to be ever present and ever active in a thinking classroom. After you have given the tasks, spend some time watching to make sure that groups have understood the instructions and are proceeding as intended. If a group is way off track, go to them and work with them to get back on the right track. This may involve restating the task for them or directing them to an adjacent group that can help redirect them.

Q What if a task I am giving them involves some technique or procedure that they do not yet know? What do I do?

A The short answer is you give them the procedure. But you have only three to five minutes to do so. This will be discussed more in Chapter 9. For the time being think hard about what the minimum knowledge that is necessary for them to start the first task and what can they learn in the first task that will help them with the second task, et cetera.

Q In this chapter it was mentioned that we would be correctly giving tasks verbally if a student walks in late and, based on what is written on the board, does not know what is going on. That feels wrong to me.

A The goal is not that students don't know what is going on. We don't want to be unnecessarily obtuse with our instructions. The idea that a student who comes to class late cannot decode the textual residue on the board and know what to do is an indicator that your instructions were mostly verbal. As mentioned in Chapter 2, if a student does come late, plug them into a group of two and have the group explain what the task is.

Q How do I distinguish between the parts that I should say and the parts that I should write?

A An easy rule to follow is to say words, and write numbers, symbols, and images. The exception to this rule is that you can also write names, labels, or modifiers such as *height, area, first, last,* et cetera. In essence, you write words that help the student demarcate information from other information. For example, if you are giving a task that involves two different speeds, you may write "speed of the bird = 4 m/s" and "speed of the ball = 5 m/s." The more information the task has, the more demarcation is needed. If there is only one number in the task and it is the speed of the bird, simply writing "5" or "5 m/s" tends to be enough.

Q I am currently doing a data analysis (or graphing) unit. Can I write the data (or graph) on the board, or should I be saying it?

A The data (or graph) are details about the task that you do not want to burden the students' cognitive loads with. Many teachers will provide each group with a sheet of paper that has the data (or graph) on it. Then, what to do with the data (or graph) is given verbally.

Q I work in a reality where students have to take externally set final exams that have a heavy focus on textual tasks. How do verbal instructions help prepare them for that?

A They don't. Verbal instructions fast-track thinking. Toward the end of the year, you can start to prepare your students for textual tasks. Some teachers will do this by once in a while giving a task textually and having the groups spend 5–10 minutes decoding what they think the task is, asking and then debriefing this in a whole-class discussion. This gradually gives way to having them decode and solve in groups before discussing the task as a class, and so on until students are doing it individually. But know that the teachers who do this well do not begin with this until the last third of the year, and then they do it only intermittently.

Q I do preteach my students how to do tasks, and I do it so they can be successful. What is wrong with that?

A There is nothing wrong with wanting your students to be successful. I think we all want that. The question is really what it means to be successful. For a long time, education in general has taken that to mean that they can mimic well. Out of this has emerged a discourse of being deliberate and intentional about the examples and instructions we give students. But, as discussed previously, mimicking is not the same as learning, and mimicking is antithetical to thinking. So, if we want our measure of success to be that our students are thinking, then we have to be deliberate and intentional about how we create and maintain an environment that promotes and sustains thinking—and this cannot include mimicking.

Q Somehow, this chapter does not feel relevant to my kindergarten classroom. We already use verbal instructions.

A Correct. The part about verbal instructions is redundant for primary teachers. However, where and when we give the task is still relevant. We found that, even in primary classrooms, there was a decrease in the number of questions asked when students were given the task early in a lesson versus later in the lesson. The same was true of having students stand versus sit. However, this does not invalidate practices like carpet time at the beginning of the day, which take well over five minutes. The research showed that it was important to give the tasks within three to five minutes of when the teacher declares that the lesson has started. In primary grades, this declaration can come after carpet time—"OK. We are now going to do a new activity. So, I want everyone to stand up and meet in that corner over there, and I'll give you the instructions."

Q How important is storytelling in giving tasks and instructions verbally?

A In every case in which we were able to create a story, students' uptake of the task was better—they had fewer questions, they were able to more quickly begin the task, and they were less likely to misunderstand what they were meant to do. There are loads of research that support these observations (Egan, 1988) as well as research on how to teach mathematics through storytelling (Zazkis & Liljedahl, 2008), but not all tasks lend themselves to being posed as a story, and not all teachers want to be storytellers. What is inherent in storytelling that can transfer to all tasks, however, is a narrative structure—a sense of chronology. The palindrome task does not lend itself to storytelling. That did not prevent us from creating a scripted narrative wherein the task emerges out of discussion and dialogue with the students.

So the tax man...

 SUMMARY

MACRO·MOVES

☐ Give the first thinking task in the first 3-5 minutes after you begin the lesson.

☐ Give the thinking task with the students standing loosely clustered around you.

☐ Give the instructions and thinking task verbally.

MICRO·MOVES

☐ Identify and create locations in the room where there is enough space for all the students to stand comfortably.

☐ Try to use different locations around the room for presenting tasks.

....MICRO·MOVES

☐ If new knowledge is needed to do the first task, think about what the minimum new knowledge needed is, as well as the minimum things that need to be said and written to pass on that knowledge.

☐ When giving a task, write on the board only the details that the students would otherwise need to remember - quantities, measurements, geometric shapes, data, long algebraic expressions, etc.

☐ When the students have started working, ask yourself if what is written on the board would make sense to a student who comes in late.

QUESTIONS TO THINK ABOUT

1. What are some of the things in this chapter that immediately feel correct?

2. Think about your teaching when students are sitting in their seats. How many are really paying attention to you? If a teacher were standing in the back of your class and was able to see what your students were really doing, what do you think they would see?

3. What is it about the students standing in close proximity to you that changes the way they pay attention to you?

4. Think about how often you are verbal, only verbal, in your current practice. Don't just think about when you are talking to the class as a whole, but also when you are interacting with the students one-on-one or one-on-few.

5. Think about how often you are verbal in your interactions with people outside of the classroom. What are the circumstances in which being verbal is not enough, and you need to demonstrate, point, or write something to help with the interaction? What is it you show, point to, or write in those circumstances? How does that compare to what you write for students in your current practice?

6. What will be the hardest part of trying to be verbal when giving a task?

7. What are some of the challenges you anticipate you will experience in implementing the strategies suggested in this chapter? What are some of the ways to overcome these?

☑ TRY THIS

The following tasks are ideally suited for giving verbally according to the scripts provided.

Grades K–3: Next Door Numbers

Teacher Let's look at these boxes. How many boxes are there?

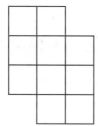

Students 10!

Teachers Correct! Now look at this list of numbers. How many are there?

1, 2, 3, 4, 5, 6, 7, 8, 9, 10

Students 10!

Teacher What your job is, is to place the 10 numbers into the 10 boxes. But there is one rule. Two numbers that are next to each other in the list, cannot be next to each other when they are in the boxes.

[Teacher makes random groups.]

Some students will interpret "next to" to mean side by side. This is OK. When they are done arranging the numbers in the boxes, snap your fingers and say, "Right. Numbers that are next to each other in the list also cannot be above or below each other when in the boxes." When they are done with whatever rearrangement this rule requires, snap your fingers and say, "Right. Numbers that are next to each other in the list also cannot be corner-to-corner with each other when in the boxes."

Source: Adapted from the Next Door Numbers task by © Crown Copyright 2000.

Grades 4–12: Tax Collector

Give the tax collector task as scripted in this chapter. When a group has maximized their gain for 12 envelopes, go to 18, then 24, then 30 envelopes.

Source: Adapted from Diane Resek task "The Tax Collector" (2007).

CHAPTER 7

WHAT HOMEWORK LOOKS LIKE IN A THINKING CLASSROOM

Of the hundreds of teachers whose classrooms I have visited and the thousands of teachers I have worked with, almost all of them give homework. And almost all of them struggle with its effectiveness. In this chapter, you will read about the role that homework plays in mathematics classrooms, how students engage with it, and how this aligns with the intentions of the teacher. Along the way you will be introduced to the limitations of the current institutionalized notion of homework and by the end of the chapter learn about an alternative conception that puts thinking and student responsibility back into homework.

THE ISSUE

Homework, in its current institutionalized normative form as daily iterative practice to be done at home, doesn't work. Almost every teacher I have interviewed says the same thing—the students who need to do their homework don't, and the ones who do their homework are the ones who don't really need to do it. It is a broken construct that long ago lost the good intention under which it was conceived.

If I ask a *teacher* what homework is for, I almost always get the same answer—homework is a chance for students to test their understanding, to learn from their mistakes, and to find what they need more help with. These are lofty objectives. When I ask *students* what homework is for, they tell me that it is for marks. When I ask them *who* homework is for, they tell me it is for the teacher. And when I ask them *why* their teacher gives them homework, they tell me it is for practice.

There is a huge disconnect between what teachers and students see as the objectives of homework. What is causing this disconnect, and what are the implications for teaching and learning mathematics in general, and for the thinking classroom in particular?

> There is a huge disconnect between what teachers and students see as the objectives of homework.

THE PROBLEM

Before beginning to experiment with alternative forms of homework, we decided to first take a very close look at how students were engaging with more institutionally normative forms of homework. As we did with the research on studenting behaviors for now-you-try-one tasks that was presented in the introduction, we wanted to know what the spectrum of studenting behaviors was for homework. So, we spent time talking to and interviewing students in classrooms where teachers gave homework as a list of questions to be done at home. In some of these classrooms, homework was worth marks, and in some it was not. What emerged from this research was a set of four basic studenting behaviors—*didn't do it, cheated, got help,* and *tried it on their own*—within each of which there were nuances and variation (Liljedahl & Allan, 2013a).

Didn't Do It

 Students don't do their homework for four basic reasons, one of which is that they don't have time. In some of these cases, homework is displaced by things like hanging out with friends, playing video games, social media, or streaming shows. But often, homework is displaced by more lofty pursuits like sports, volunteerism, work, family functions, music, and other homework. From the interviews, we learned that a lot of students are legitimately very busy. Many younger students are participating in multiple sports, and older students have multiple teachers who are giving homework.

A second major reason students don't do homework is that they forget. This forgetfulness is usually symptomatic of two issues—homework is not important to them, and/or they have poor record keeping. Usually students who are forgetful are not *always* forgetful and do their homework sporadically. A more common reason students do not do their homework, however, is that they don't know how to do it. This is often masked by excuses of being busy or forgetting. At its core, not knowing how to do the homework is the most legitimate reason for not doing it. Finally, students do not do their homework because they don't want to do it. If it is not worth marks, it is easier not to do it. Even if homework is randomly checked, students may take a chance that this particular homework assignment will not be marked and choose not to do it.

Not doing their homework was not something that was unique to middle and high school students. This behavior was seen with almost equal distribution in all classes where homework was assigned—irrespective of grade.

Cheated

If you ever want to have a good day, ask some middle or high school students to tell you about some of the ways in which they cheat in other teachers' classes. Some students are amazingly innovative and industrious and are often keen to share some of their brilliance when it comes to cheating. When it comes to homework, cheating predictably includes copying from someone else or borrowing someone else's work. But their ingenuity goes well beyond these tried and true methods. A number of students we interviewed admitted that they had access to an entire binder of worked homework solutions from students who had taken the same course with the same teacher in the past—something that only works with teachers who use the same homework questions year after year.

Even more industriously, several students had pages in their binder ready to go if there was ever a homework check. These pages were covered in dense mathematics, some had diagrams, and some had graphs. When the teacher began to walk up and down the aisles checking that students had done their homework, these students would display some of these pages in lieu of the actual homework. We were told by these students that this almost always works. And if it didn't, they had an out. If the teacher noticed that it was the wrong homework, something they assured us rarely happened, they would just say, "It is?" and then start randomly flipping back and forth in their binder until the teacher became impatient and walked away. Unbeknownst to us, we witnessed this working twice. Only after the homework check had happened did students reveal to us what they had really done.

The most innovative form of cheating we witnessed, however, were students who had the same course with the same teacher, but in different blocks, who shared a binder or a workbook. These students had brilliant protocols for division of labor that went well beyond homework and included note taking and test review.

Regardless of the method of cheating, however, any student who admitted to cheating was asked to explain why they did it. The least common, but most interesting, answer was that some students cheated because it was fun—they liked the subterfuge and the excitement of it. More common was that they didn't know how to do the homework. Cheating, for these students, was a way to mask their lack of ability. The most common reason for cheating, however, was that the homework was for marks—and cheating assured that the marks would be gotten. Sometimes the reason for cheating was a combination of two or three of these reasons.

Although cheating was a behavior that we saw more with Grade 6–12 students, there were also some cases of cheating in students as young as Grade 2. In these cases, the most common form of cheating was copying from a peer, which was something peers were very reluctant to allow. When talking to the students who cheated, it became clear that they did this primarily because they forgot to do their homework and they wanted to avoid disappointing their teacher.

Got Help

Students often get help with their studies. From homework to preparing for tests, some students routinely seek help from peers, parents, tutors, and teachers. This was neither unexpected nor uncommon. What was interesting, though, was the reason behind, and the results of, getting help.

> When students who got help from a tutor or parent were asked how they would do if a pop quiz based on the homework were given, 90% of the students said they would fail.

When asked why they sought help, almost all students claimed it was because they didn't know how to do the assigned work. There were a few cases where working with peers, parents, tutors, or their teacher was part of the daily homework routine irrespective of help being needed or not. When students who got help from a tutor or parent were asked how they would do if a pop quiz based on the homework were given, 90% of the students said they would fail. So, what did they get help with? They got help with getting the homework done—not with learning.

Drilling deeper into these data, it turns out that the exceptions were the students who worked with adults as part of their homework routine, whether help was needed or not. These routines seem to be responsible for the lower number of elementary and middle school students who said they would fail such a quiz. For them, the likelihood that they have a homework routine with their parents was much higher, and hence their results were better. The other exception was the students who had homework routines that involved working with peers, all of whom said they would pass.

Tried It on Their Own

The rest of the students tried the homework on their own. Some completed it and some did not. Regardless, they did not say they forgot, cheated, or got help. Of these students, the vast majority completed the homework by mimicking from either their notes or the textbook. Like the students who mimicked on the now-you-try-one tasks discussed in the introduction, when asked about this, all of these students said they thought this is what they were supposed to be doing.

Lukas   Why else would he have notes?

Fatima   Isn't that what the teacher wants us to do?

Samantha   The teacher shows us how to do it and then we need to practice it. Right?

In fact, this mimicking behavior was such a dominant strategy that when the examples ran out, so did their ability to answer homework questions.

Researcher Were you able to do all of the question?

Stephan   Yup. Except for the last two. I didn't have examples for those.

Of all of the students we interviewed who used mimicking as a strategy to complete the homework on their own, less than 20% were even willing to move beyond this strategy when the examples ran out, and less than half of those were able to answer questions for which an analog did not exist in their notes or the textbook.

Among two hundred students in Grades 4–12 that were interviewed, these four behaviors—not doing it, cheating, getting help, and doing it on their own—were distributed almost evenly in classrooms where homework was marked (Figure 7.1). When homework was not marked, cheating disappeared almost completely (Figure 7.2), and, whereas the number of students who did not do their homework increased, so too did the number of students who did it on their own. That is, in situations where homework was marked, approximately 50% didn't do it or cheated. When homework was not marked, this percentage drops to approximately 40%—not marking homework had a positive effect on how many students did their homework.

> Not marking homework had a positive effect on how many students did their homework.

Of course, these numbers vary between elementary and secondary students, with more elementary students getting help than secondary students, and fewer forgetting or cheating. There doesn't seem to be any sort of linear deterioration in the way homework is approached as students get older, but it is clear that when marks start to matter to the students and their parents, their homework behavior changes markedly for the worse.

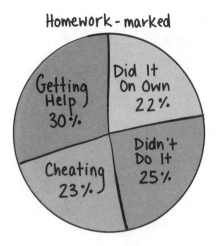

Figure 7.1 Studenting behaviors when homework is marked.

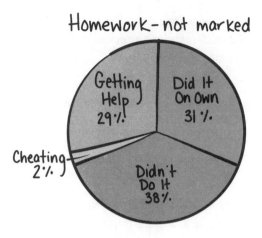

Figure 7.2 Studenting behaviors when homework is not marked.

> Homework, in its current formulation, needs to be upgraded. It needs to be rebranded. It needs to become a thinking activity.

These data confirm what we already know—what you likely already know. Homework is not working. Students are doing it, if at all, for the wrong reason (marks) and the wrong person (their teacher or their parents). And those who are doing it for the right reasons (to check their understanding) and for the right person (for themselves) are mimicking. Homework, in its current formulation, needs to be upgraded. It needs to be rebranded. It needs to become a thinking activity.

TOWARD A THINKING CLASSROOM

In the history of schooling, homework has been rebranded once before—as *practice*. In many classrooms, especially elementary classrooms, the word *practice* is still used either to describe work done in class or used synonymously with homework. Regardless, practice puts a greater emphasis on mimicking while still not resolving the issue of why it is done and who it is done for. As Samantha's comment (above) shows, when the terminology of practice was used, it increased the perception that mimicking was what students were meant to be doing and, as a result, increased mimicking behavior. And, as discussed, mimicking has limitations and is antithetical to the kind of thinking behaviors that thinking classrooms are trying to foster.

So how do we rebrand? Let us come back to the notion of why homework is done and who it is done for. We've seen that if the purpose of homework is to get marks and/or if it is done for the teacher (or parents), it loses its potential to achieve what teachers really want it to achieve—for homework to be a safe place for students to make mistakes as they check their understanding. In here lies the rebranding needed. So, we stopped calling it homework and started calling it *check-your-understanding questions*.

> We stopped calling it homework and started calling it *check-your-understanding questions*. This had an immediate effect on students.

Calling it check-your-understanding questions specified who it was for—the student (*you*)—and what it was for—*to check understanding*. This had an immediate effect on students. We saw more students doing check-your-understanding questions on their own than had we seen with "homework"—and they were doing it for the right reason. Even when students were seeking help, they were now seeking to understand rather than seeking to be done. And, students were, for the most part, no longer talking about marks, practice, mimicking, or doing it for the teacher or their parents.

Of course, it was not as simple as just relabeling. As with the other practices discussed thus far, the use of check-your-understanding questions needed to be accompanied by a slate of other changes to teaching practice. First, the questions could not be marked. They couldn't even be checked. In fact, there can be no overt actions on the part of the teacher to enforce that the questions are being done—either positively or punitively. Any efforts to do so were met with an immediate and almost complete transition back toward these questions being done for the teachers. In essence, if you want check-your-understanding questions to be a safe place for students to make mistakes, then you have to keep it safe.

This was difficult for us to accept. In every case where we implemented check-your-understanding questions, 15%–50% (usually 15%–25%) of students didn't do them. This sounds bad. But on the flip side, this meant that 50%–85% (usually 75%–85%) of students were doing the questions, and doing them for the right reason and for the right person. Although the studenting data on homework showed 75% of students completing their homework, only about 10% were doing so for the right reason. When completion is the goal, it encourages, and sometimes rewards, behaviors such as cheating, mimicking, and getting unhelpful help.

This is not to say that you cannot talk to your students about them doing their check-their-understanding questions. This turned out to be very important—but also very risky. The discourse around this, as it turns out, needs to be focused on check-your-understanding questions as an opportunity to learn from their mistakes (without risk), to check their understanding, and, above else, is for them and only them. We need to drop any references to words like *practice*—which invokes mimicking behavior—and *assignments*—which invokes a sense that it is for a mark.

Another change in practice was that answers needed to be provided at the same time as the questions were given. If check-your-understanding questions were truly to be seen as a way for students to check their understanding, they needed something to check against—they needed answers to see if their understanding was correct. Fully worked out solutions can also be provided, but not right away. We learned that when students see these before they have done the questions, some mistake their understanding of the worked solution for an ability to do the questions on their own and choose not to do them.

So, the questions may be the same as what we previously gave as "homework." But to make doing them a thinking activity, everything around them changes—what we call it, how we talk about it, the autonomy students have to do it, and our openness to the fact that students may not do some or all of the questions. And when these changes occur, the doing of check-your-understanding questions becomes a thinking activity (see Figure 7.3).

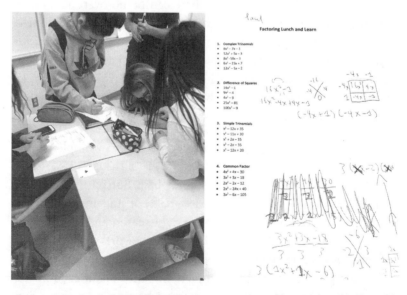

Figure 7.3 Students doing check-your-understanding questions at lunch.

FAQ

Q If I don't check or mark the check-your-understanding questions, most of my students will not do them. Are there any strategies I can use to get them to do them?

A Let's assume that we have half of your students doing the check-your-understanding questions, and doing them for the right reason. First of all, these are already better statistics than we saw in the studenting data around homework. Second, the research showed that if we start policing these questions, we will be successful in raising the number of students who do them. But we will sharply decrease the number of students who are now doing them for the right reason—for themselves to check their understanding. Of all the things we researched, check-your-understanding questions is one of the most sensitive to disruption. This is not to say that there are not small things you can do to encourage them to do these questions for themselves. As mentioned, our discourse around this practice is helpful. The use of words such as *opportunity* are helpful—and the use of words like *practice* and *assignment* are not. If you are doing this within the context of the other thinking classroom practices discussed so far, talk to the students about how check-your-understanding questions is a chance for them to see if they can do on their own what they already did in their groups. In essence, if you can keep the focus on them and the correct reason for doing the questions, anything you say is helpful.

For example, one teacher I worked with asked their students to discuss amongst themselves which of the check-your-understanding questions they thought were the most important for them to do. This metacognitive discussion had a significant impact on how many students did at least some of the questions.

Q I don't give homework. I give my students questions to do in class, but I don't send anything home for them to do. Should I still be changing what I call these questions done in class to check-your-understanding questions?

A Yes. Unlike *homework*, the name *check-your-understanding questions* does not specify where and when they are to be done. Neither does it specify that they must be done. This is about messaging around why and for whom these questions are done. Having said that, the research showed that as long as we're not policing, giving class time to do these questions increases the number of students who do the questions and the number of questions they do.

Q So much of building thinking classrooms is about collaboration. Can the students do their check-your-understanding questions in groups?

A Check-your-understanding questions are for the students, and they have autonomy over all aspects of them. If they choose to do them in groups, we cannot and should not prevent them from doing so—and we should not control who is in the groups. In fact, if you give class time to do check-your-understanding questions, 40–70% of your students will choose to do them in self-selected groups of two or three. Many of these groups will choose to do them on VNPSs, and many will stay in the random groups from the beginning of the lesson.

Although they are working collaboratively, what is interesting is that the goal remains the same—for them to check their own understanding. And with this comes a very interesting discourse as they strive to learn from each other—"I don't get it. . . . I still don't get it. . . . Ok. I think I get it. Give me another one to see if I got it." Although they are working together, the goal remains fixed on checking their understanding.

Q So, you say that we can give worked solutions—but not right away. Should we do this? And if so, when?

A First, to make sure we are all on the same page, we need to differentiate between answers and worked solutions. $x = 7$ is an answer. It reveals nothing about how we arrived at that solution. How we arrived at that solution is the worked solution. The research showed that giving worked solutions became more and more important the more complex the questions were. This corresponds loosely with an increase in grade level, but not exactly. The research also showed that worked solutions should not be given out until the students have had a chance to work on the questions, checked the answer they arrived at against the answers provided, and, if needed, retried the questions—sometimes multiple times. As mentioned above, if we give the worked solution at the same time as the question, some students will read the solution and think that, because they understand what is happening, they understand how to do it on their own. However, if the worked solutions are provided the next day, or a few days after, the students now have a chance to compare their thinking to that of the worked solution as well as to get help with any questions they were unable to resolve on their own. Many teachers I worked with gave out worked solutions by posting them on a class website or portal.

Q I have parents asking for homework or practice. How do I deal with this?

A Begin by communicating with them about why you give "homework" and how these reasons are embodied in the name *check-your-understanding questions*. Explain that the questions that may or may not be coming home are to be treated as a safe place for their child to make mistakes and that the purpose is for them to learn from these mistakes. If they insist on having more questions for practice, you can tell them that ample practice questions can be found on the internet, but that they will not be part of your teaching practice.

Q I teach primary grades, and I do not typically assign homework. Should I be doing check-your-understanding questions?

A Yes. But I am willing to bet that you already do. Every teacher I have ever worked with from kindergarten through Grade 12 gives students questions to do. Irrespective of what you may call these questions, you are likely giving them as a way for students (and you) to check their understanding.

SUMMARY

MACRO·MOVE

☐ Give students an opportunity to do check·your· understanding questions.

MICRO · MOVES

☐ Do not mark it.
☐ Do not check it.
☐ Do not ask about it.
☐ Don't use words like PRACTICE or ASSIGNMENT.
☐ Use phrases like THIS IS YOUR OPPORTUNITY.

> I don't get it
>
> I still don't get it!
>
> OK. I think I get it.
>
> Give me another one to see if I got it.

☐ Provide answers at the same time when you give check·your· understanding questions.
☐ Provide worked solutions a day (or so) after giving check· your·understanding questions.
☐ Give students a chance to discuss which questions they think are important for everyone to do.

QUESTIONS TO THINK ABOUT

1. What are some of the things in this chapter that immediately feel correct?

2. Which of your students do their homework, and which do not?

3. Of those who do their homework, why do they do it? If you consider yourself successful at getting students to do their homework, what message are your methods sending to your students? That is, why do they do their homework, and who is it for?

4. Contrast your answers to Question 3 with the reasons for why *you* want your students to do their homework.

5. In this chapter, it was mentioned that practice invokes mimicking. What are your thoughts about practice as an effective learning tool? Is this what you want your students to do?

6. What do you think about the reality that some students may choose not to do, or not do all of, the check-your-understanding questions? How will you cope with this?

7. What are some of the challenges you anticipate you will experience in implementing the strategies suggested in this chapter? What are some of the ways to overcome these?

☑ TRY THIS

Rebrand what you would previously refer to as homework or practice into a set of check-your-understanding questions. Give it to your students with the answers, and emphasize how this is an opportunity for them to see whether they have understood what happened in their groups. Do not collect it, or mark it, and the next day provide the worked solutions for them to use to check whether they truly did understand it.

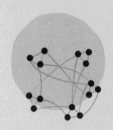

CHAPTER 8

HOW WE FOSTER STUDENT AUTONOMY IN A THINKING CLASSROOM

· ·

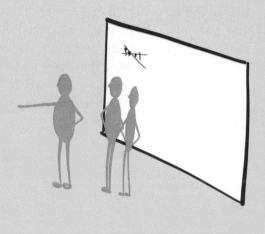

Whether or not you have been implementing the practices as you have been reading along, by now you will have concluded that a thinking classroom looks very different from a typical classroom. Students are working in groups rather than individually, they are standing rather than sitting, and the furniture is arranged so as to *defront* the room. Closer inspection will reveal that the teacher is giving instructions verbally, is answering fewer questions, and has drastically altered the way they give "homework." All of these changes require a greater independence on the part of the students. This chapter will look at how you can begin to build this independence through giving students more autonomy over their own actions and how this autonomy not only changes the way students engage with a thinking classroom, but also how it makes your job as a teacher in a thinking classroom easier.

> All of these changes require a greater independence on the part of the students.

Figure 8.1 Students working independently in random groups on VNPSs.
Source: Used with permission

? ⦶ ? THE ISSUE

From a teaching perspective, one of the big differences you may be experiencing, or anticipate experiencing, is that in a thinking classroom you, as the teacher, have a lot less control over what is happening in the room. When students are sitting in desks, all facing the front, and presumably following along with what is happening at the board, the teacher has a lot of control around what is happening for all the students in each moment—in this moment all the students are doing notes, in this moment all the students are doing now-you-try-one questions, in this moment all the students are starting their homework, et cetera. Everything is sequenced, paced, and synchronized so that everyone is doing the same thing at the same time—or at least seemingly so. We know from the *studenting* research presented so far that what goes on behind this façade is quite different from what we intend and, even, what we see. Regardless, there is structure. And the stronger the structure, the lesser the need for the students to be independent—and the lesser the need for students to have autonomy.

THE PROBLEM

Lack of autonomy is synonymous with lack of choice. And lack of choice reduces the need for students to think. The research clearly showed this. The amount of thinking students were required to do, and did, was sharply reduced in situations where their actions were managed—even micromanaged. In thinking classrooms, students' actions cannot be managed the same way as in a typical classroom. When students are working in random groups on vertical non-permanent surfaces, you can typically only work with one group at a time. This means that at any given moment there are 8–10 groups that need to be working independently. There is much happening outside of your control, and in order for it all to work well, students need to take on much more responsibility for their learning. This cannot happen unless they have the autonomy to do so. The question is, exactly what should they have autonomy over, and how are you going to foster this?

> The amount of thinking students were required to do, and did, was sharply reduced in situations where their actions were managed—even micromanaged.

TOWARD A THINKING CLASSROOM

Early in the research into building thinking classrooms, it became apparent that every teacher implemented the thinking classroom practices slightly differently. Whereas some teachers used cards to randomize their students, others would use a computer app. While some teachers had students working on whiteboards, others had them working on windows or tables standing on end. Where some teachers gave class time for check-your-understanding questions, others asked for them to be done at home. And so on. This was totally OK, as these practices are meant to be thought of as a framework much more than an overly prescriptive set of pedagogical moves that must be adhered to. As mentioned in the introduction, the building thinking classrooms framework is a collection of empirical results that offer teachers a chance to fundamentally change the way students experience mathematics. An individual teacher may not wish to, or be able to, implement each practice in the exact way that the research showed was best. And even if they do, they have to find a way to personalize it and make it their own practice. That is, each of you has autonomy over whether, and how, you implement the thinking classroom practices.

For me, the differences I was seeing from room to room were fascinating, and I wanted to learn more from them. One of the biggest differences was how many students' hands were raised, and how often. In some classes many students had their hands up every time they were done with a question or when they were stuck. In other classes I saw no hands go up—yet, these students seemed to progress just as far, even further, than those in classes where the teacher was providing a lot of help. Just to be clear, this research was done prior to the research on how to, or how not to, answer students' questions.

When I began to pay closer attention to the classes where students rarely, if at all, put up their hands, I began to notice that there was much more interaction between groups and that this interaction was both passive (looking at other groups' work) and active (talking to other groups). Further, I noticed that these interactions occurred most often at moments when a group was either finished or stuck. That is, when a group was finished with their current task, they would look around the room to first confirm their answer and then look for a question that another group was working on that they had not yet solved. If, in their efforts to confirm their answer, they saw conflicting answers, they would either take this as a cue to dive back into their own work or go and talk to the group with the different answer.

> There was much more interaction between groups, and this interaction was both passive (looking at other groups' work) and active (talking to other groups).

If students were stuck, they would behave in much the same way and either passively look at other students' work to see if this could inspire an idea of how to proceed or hint at what they may be doing wrong. Often these passive looks resulted in the *borrowing* of notation and/or organizational tools such tables, graphs, diagrams, et cetera. But sometimes it would prompt a more active interaction in the form of a conversation with another group.

What was clear from all these interactions was that knowledge moved easily from group to group. This made the work of the teacher much easier.

What was clear from all these interactions, both passive and active, was that, in these classrooms, knowledge moved easily from group to group. This made the work of the teacher much easier. The more independent and responsible the students were in managing their learning, the more focused time the teacher had to spend with groups that really needed their attention. It was also clear that this independence was a product of the autonomy the students were given to both passively and actively interact with other groups and their ideas within the room—to mobilize the knowledge in the room.

This is not to say that the students had total freedom to do what they wanted. If students from different groups were meeting up to discuss something not related to mathematics, the teachers were quick to get them back on task. The same was true if the teachers saw students using their cell phones for non–mathematics related activity. The autonomy the students were afforded was specific to knowledge mobility. If you need help, get it. If you need another question to work on, find it. This was clearly important and, therefore, became one of the practices identified as being needed for building thinking classrooms.

Ironically, when I mentioned autonomy to the teachers who had many students who frequently put up their hands, all of them said that they do give their students the autonomy to look around and interact with other groups and that they often tell their students that it is OK to do so. I even watched three of these teachers do just that, to no effect. So, I went back to the teachers whose students were not only given autonomy but who also used that autonomy to keep themselves moving forward through passive and active interactions. Even closer inspection of what the teachers did in these classrooms revealed that they, themselves, frequently made use of the knowledge in the room to move groups forward. That is, rather than directly answer questions, help, or give the next question, they would sometimes direct a group's attention to what another group was doing.

Teacher   So, where are you at with this?

Group   I think we're stuck. We're just going around and around.

Teacher	Hmmm . . . Why don't you look at how that group [teacher points at Group 3] has organized their data? That might help.
Teacher	What's going on? You guys are just standing here.
Group	We're done. What's next?
Teacher	Why don't you ask the group next to you? [Teacher points at Group 8 on their right.] They seem to be working on something different.

These types of moves were seen much less frequently in the classrooms where students put up their hands often. In those settings the teacher was much more likely to suggest using a table or give the next question, respectively.

These deflective moves became something we began to experiment with. Working with teachers from kindergarten to Grade 12, we began to have teachers direct students to other groups whenever they were stuck or in need of a new question. In short, rather than being the source of knowledge in the room, teachers were working to mobilize the knowledge already in the room. That is, they were being *deliberately less helpful*. This proved to be extremely effective at getting students to seize the autonomy they had been given and begin to use it to keep themselves moving forward. After the first two-week cycle, we were seeing huge improvements in this regard, and after a second two-week cycle classrooms were, for the most part, functioning the way we wanted—fewer hands in the air and more passive and active interaction between groups.

> Rather than being the source of knowledge in the room, teachers were working to mobilize the knowledge already in the room.

Figure 8.2 Students actively and passively interact between groups to further their learning.

In the second cycle, we also began to mobilize knowledge even before groups were stuck or in need of a new question. The most obvious time to do this was when two or more groups had different answers. Getting these groups to talk to each other without specifying which was correct or incorrect (sometimes both) proved to be a very effective way to deepen students' thinking.

Teacher So, this group over here has 45 as their answer [pointing at Group 4]. And this group has 51 as their answer [pointing at Group 5]. I can guarantee you that at least one of you is wrong. But, I am willing to bet there are parts of your answers that you both agree on. Start talking about the parts you agree on, and then talk about the parts that you do not.

The same was true when groups had the same answers but different ways of approaching them.

Teacher I see that both of you have the same answer [pointing at groups 7 and 8]. But you did it in completely different ways. I would like each group to try to figure out what the other group was thinking.

> Purposefully mobilizing knowledge by forcing either passive or active interaction between groups activated the autonomy students had been given, increased independence, and deepened thinking.

Purposefully mobilizing knowledge by forcing either passive or active interaction between groups activated the autonomy students had been given, increased independence, and deepened thinking. It also increases the porosity between groups that was started with the random groupings discussed in Chapter 2.

Activating their autonomy also had an effect on how students thought about their work and their workspace. First, groups stopped trying to shield their work from others, and comments such as "they're cheating" quickly disappeared. On the other hand, students were much more likely to look around without feeling like they, themselves, *were cheating*. And there was an increased sense that they all had something to offer others.

Group 1 The teacher sent us over here to see what you guys are doing.

Group 2 That must mean that one of us has something to share. I wonder if it is you guys or us?

Autonomy, like room organization, was not something that I began my research thinking about as an independent variable in need of research. But, as it did with classroom organization, spending time in classrooms and experimenting with practices that increase thinking revealed that autonomy is a variable that needs attention.

A thinking classroom is a classroom where students think individually and collectively. The collective goes well beyond the limits of the group boundaries and encompasses the whole class. We need to give groups the autonomy to make use of the knowledge in the room. But this, it turned out, was not enough. We need to also help them to break down the barriers around their groups by mobilizing the knowledge in the room for them. Not only does this build the independence that is needed for a thinking classroom to function well, it also engenders the type of 21st century skills that people need to work and collaborate in the real world.

> We need to give groups the autonomy to make use of the knowledge in the room. We need to also help them to break down the barriers around their groups by mobilizing the knowledge in the room for them.

FAQ

Q So, does this mean I should stop answering all questions?

A No, it doesn't. But if you want your students to begin to seize the autonomy you have given them, then look for moments where a group can get what they need from another group. When you find such a moment, the key is to not say or show something another group can.

Q If I give and foster this kind of autonomy, won't some groups just begin to copy from other groups?

A As mentioned in the FAQ of Chapters 2 and 3, this actually turns out to be a very rare occurrence. In the hundreds of times I have been in a thinking classroom, I have never seen a group copying, line by line, from another group. By and large, when a group looks at another vertical surface, they look for hints. These hints often take the form of notation, organization, or key ideas. Once they have acquired the hint, they go back and solve the problem for themselves. Or they're looking for validation, so that they feel confident to move on. Eventually, students start to look to other groups for extensions. In short, groups look for the things that you would feel comfortable giving them if you were to help them.

Q Is there ever a time where I would want a group to work entirely on their own without any hints from the other groups in the room?

A There are two instances where you may wish for this. The first is if you ever wanted students to answer some check-their-understanding questions where you want them to really see whether they understand what they are capable of doing entirely on their own. The second we will see in Chapter 14, when we discuss the possibility of doing group quizzes. In neither of these instances is it necessary that groups work without interactions with other groups, but if you wish for this to be the case, you can give every group a different question. This is more taxing on you, but it achieves what you are looking for.

Q Is it helpful to put two groups together that have the same strategies or the same solutions?

A Yes. Every group develops their own way of talking about a problem and their own way of representing it. This is why, when you put two groups together, one of the first things they do is to negotiate language and notation. So, even if you put two groups together who have seemingly similar solutions, these negotiations will strengthen their understanding of the solution.

Q How can I tell whether groups are actively interacting for the purpose of mobilizing knowledge or just for socializing?

A The short answer is to listen to them. However, if you are not close enough to listen, then there are a few visual cues you can attend to. First, if one member of a group is talking to only one member of another group, then there is a strong chance that they are socializing. If both groups have backed away from the vertical surfaces and no one is gesturing toward the vertical surfaces, there is also a strong chance they are socializing. However, keep in mind that I have seen both of these occur and the discussion being 100% about the mathematics and the problem at hand. Also keep in mind that when two groups first begin to actively interact, they may begin the interaction by joking and talking about things off topic. This is a normal part of social interaction and should be tolerated. If it goes on for too long however, you need to intervene to get them back on task—*"OK. Now that everyone has gotten to know each other, how about we get on with the math?"*

Q If one group is helping another group, how do I know that they are helping and not just telling them how to do it?

A For the most part, it doesn't matter. Regardless of how a group acquires new knowledge, they tend to take that knowledge and try to apply it for themselves. Even if they do not, the extension they are next asked to work on often requires them to apply their new knowledge anyway. Having said that, it is always best if groups who are in the helping role are thoughtful about how they give out ideas so as to maximize the opportunity for the learning group to think. You may wish to discuss this with your class from time to time—but not until after you have begun to mobilize knowledge.

Q What if the help that a group actively gives, or a group passively gets, is mathematically incorrect?

A This happens. But, in the collaborative setting of a thinking classroom, groups tend to self-correct. After all, the more eyes that are on something, the less likely an error goes unnoticed. For this reason, mobilized knowledge, like collaboration, also tends to converge toward correctness. And do not forget that you are still in the room and monitoring progress. If you see an error, you can point it out.

SUMMARY

QUESTIONS TO THINK ABOUT

1. What are some of the things in this chapter that immediately feel correct?

2. Have you already given your students autonomy to interact across groups? Have they taken advantage of this autonomy to the degree that you wish?

3. What other ways might you foster autonomy beyond what is mentioned in this chapter?

4. This chapter focused on the nurturing of independence through the fostering of autonomy. Have you found any other ways to nurture independence?

5. What is your feeling about mobilized knowledge versus groups doing the work on their own?

6. What are your feelings about the possibility for the proliferation of errors in a classroom where knowledge is being shared between groups?

7. What are some of the challenges you anticipate you will experience in implementing the strategies suggested in this chapter? What are some of the ways to overcome these?

☑ TRY THIS

The thinking tasks that follow have been shown to produce diverse solutions and solution methods—ideal ground to foster autonomy through mobilizing knowledge.

Grades K–4: Pentominos

Using exactly five multiplex cubes, how many different shapes can you make such that there is a way to place the shape on the desk so that all five cubes touch the desk?

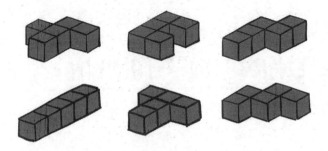

Grades 5–8: Nickels, Dimes, and Quarters

How many ways are there to make a dollar using only nickels, dimes, and quarters?

Grades 9–12: Birthday Cake

You want to arrange four candles on a cake. How many ways can you place the candles such that there are no more than two different distances between any two candles?

CHAPTER 9

HOW WE USE HINTS AND EXTENSIONS IN A THINKING CLASSROOM

1. $(x+2)(x+3) = x^2 + 5x$
2. $(\quad)(\quad) = x^2 + 7x +$
3. $(\quad)(\quad) = x^2 + 7x + 12$
4. $(\quad)(\quad) = x^2 + 14x + 24$
5. $(\quad)(\quad) = x^2 + 10x - 24$
6. $(\quad)(\quad) = x^2 + 4x - 12$
7. $(\quad)(\quad) = x^2 - x - 12$
8. $(\quad)(\quad) = x^2 - 2x - 24$
9. $(\quad)(\quad) = x^2 - 6x - 16$
10. $(\quad)(\quad) = x^2 - 0x - 16$
11. $(\quad)(\quad) = x^2 - 25$
12. $(\quad)(\quad) = x^2 - 49$
13. $(\quad)(\quad) = x^2 - 10x + 24$
14. $(\quad)(\quad) = x^2 - 13x + 12$

In Chapter 1, the claim was made that curriculum tasks can also be thinking tasks. Yet, with the exception of Chapter 7, every chapter since then has concluded with a non-curricular thinking task for you to use in trying to implement the thinking practice discussed in that chapter. In this chapter, I continue the thread started in Chapter 1 and discuss how curriculum tasks are the main staple of thinking classrooms. By the end of this chapter you will have learned not only how to craft and sequence curricular tasks for use as thinking tasks, but also how to use hints and extensions to help students think their way through these sequences and, as a result, cover large amounts of content in a short amount of time.

THE ISSUE

Mathematics teaching, since the inception of public education, has largely be been built on the idea of synchronous activity—students write the same notes at the same time, they do the same questions at the same time, they get the same hints and extensions at the same time. From a teacher's perspective, this is an efficient strategy that, on the surface, allows us to transmit large amounts of content to groups of 20–30 students at the same time.

> Mathematics teaching, since the inception of public education, has largely be been built on the idea of synchronous activity.

If we go under the surface, however, we realize that students' abilities are more different than they are alike, and the idea that they can all receive, and process, the same information at the same time is outlandish. Decades of work on differentiation is built on the realization that students learn differently and at different speeds, and have different mental constructs of the same content. What this work is telling us is that students need teaching built on the idea of asynchronous activity—activities that *meet the learner where they are* and are customized for their particular pace of learning.

> Decades of work on differentiation is built on the realization that students need teaching built on the idea of asynchronous activity.

THE PROBLEM

From a learning perspective, the notion that learners need customized attention and to go at varying paces makes sense. But from a teaching perspective, the thought of trying to do this for 20–30 students at one time can be overwhelming. Couple this with the need to move every student through large amounts of content with limited contact time, and the idea of differentiating learning for every student may seem untenable. This is not to say that it can't be done. There are scores of teachers who have effectively differentiated their instruction and provided customized learning opportunities for each of their students. Their students, in turn, are reaping the benefits of the individualized learning afforded by these teachers' efforts. Regardless, if differentiation is something that you have or have not been able to achieve in your teaching, the question remains, how does it look in a thinking classroom where students spend much of their time thinking in groups?

TOWARD A
THINKING CLASSROOM

As indicated in Chapter 1, thinking and engagement are closely associated. If we are thinking, we will be engaged. And if we are engaged, we are thinking. Although thinking is a private and invisible process, engagement has a physicality that is easily observed and easily identified. As such, engagement is a methodological tool that can be used to see when students are thinking. More than this, however, engagement is a pedagogical tool that can be used to build thinking classrooms.

> If we are thinking, we will be engaged. And if we are engaged, we are thinking.

To see this, we need to go back to the early 1970s and look at the work of Mihály Csíkszentmihályi, a Hungarian-born psychologist working at the University of Chicago. At the time, Csíkszentmihályi was very interested in understanding something that he referred to as the *optimal experience* (1990, 1996, 1998),

> a state in which people are so involved in an activity that nothing else seems to matter; the experience is so enjoyable that people will continue to do it even at great cost, for the sheer sake of doing it. (Csíkszentmihályi, 1990, p. 4)

The optimal experience is a form of intense engagement we are all familiar with. It is that moment where we are so focused and so absorbed in an activity that we lose all track of time, we are undistractible, and we are consumed by the enjoyment of the activity. As teachers, we see rare glimpses of this type of engagement in our students when the bell rings and they are reluctant to leave.

Wanting to better understand this rare and powerful phenomenon, Csíkszentmihályi studied people he thought were most likely to have experiences with it—musicians, artists, athletes, scientists, and mathematicians. Over time he gathered enough cases of the optimal experience that he could begin to look for patterns—and patterns he found (Csíkszentmihályi, 1990).

For example, he noticed that whenever someone had an optimal experience, they lost track of time, and much more time passed than the person realized. He noticed that when someone was having an optimal experience, they were undistractible and unaware of things in their environment that would otherwise interfere with their focus. He noticed that their actions became a seamless and efficient extension of their will. And he noticed that they became less self-conscious, stopped worrying about failure, and were doing the activity for the sake of doing it and not for the sake of getting done—it became an end unto itself.

These characteristics, like thinking, are all internal to the doer and, as such, invisible to the observer. However, Csíkszentmihályi also noticed that whenever there was an optimal experience, there were three qualities also present in the environment in which the optimal experience was taking place—clear goals every step of the way, immediate feedback on one's actions, and a balance between the ability of the doer and the challenge of the task. Unlike the first six characteristics, these are external to the doer and, as such, observable—and recreatable.

The third of these environmental characteristics—balance between challenge and ability—is central to Csíkszentmihályi's analysis of the optimal experience (1990, 1996, 1998) and comes into sharper focus when we consider the consequences of having an imbalance in this system. For example, if the challenge of the activity far exceeds a person's ability, they are likely to experience a feeling of frustration. Conversely, if their ability far exceeds the challenge, they are likely to experience boredom. When there is a balance in this system, a state of what Csíkszentmihályi refers to as *flow* is created (see Figure 9.1). Flow is, in brief, the term Csíkszentmihályi used to encapsulate the essence of optimal experience and the aforementioned elements into a single emotional-cognitive construct.

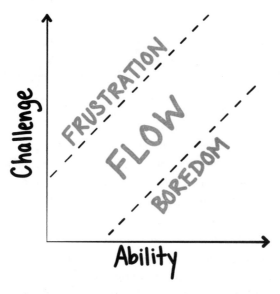

Figure 9.1 Graphical representation of the balance between challenge and skill.

Using Extensions to Maintain Flow

> To build a thinking classroom we need to be able to get students into, and keep them in, flow.

At this point Csíkszentmihályi's research on flow goes in a different direction, and my research on building thinking classrooms comes in. In essence, flow is where engagement and, as a consequence, thinking happens. Therefore, to build a thinking classroom we need to be able to get students into, and keep them in, flow. And to do this, we need to first understand how students move about inside of flow.

Representing flow as a graph that shows the balance between a doer's ability and the challenge of the task at hand invokes thoughts that this is a static space made up of fixed points, each of which represents a potential student's place on the graph. This is not how flow works. Flow is not a collection of fixed points—flow is a dynamic space. If a student works on something, their abilities will increase (see Figure 9.2), and in order to keep them from getting bored we must increase their challenge. And then their ability will increase and, eventually, the challenge will need to be increased again. And so on. In essence, when students are in flow, their ability will always be increasing, and in order to keep them in flow we, as teachers, have to keep increasing the challenge by giving them extensions—harder and harder tasks to solve.

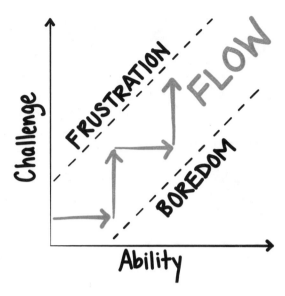

Figure 9.2 Graphical representation of the balance between challenge and skill as a dynamic process.

This sounds simple enough. And, in fact, it sounds an awful lot like what we already do. The problem is—timing matters. If we increase the challenge of a task before a student has had the chance to fully grow their ability, then, rather than keeping them in flow, we have pushed them into frustration (see Figure 9.3). Likewise, if we wait too long to increase the challenge, we push them into boredom (see Figure 9.4). The timing matters.

To keep students in flow, timing matters.

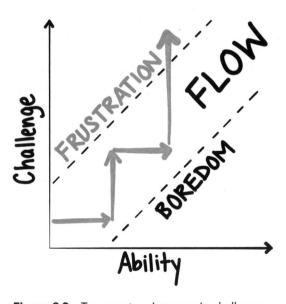

Figure 9.3 Too great an increase in challenge.

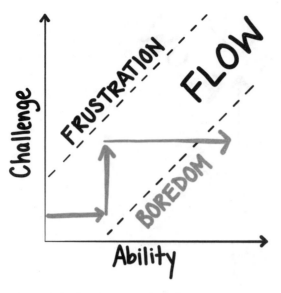

Figure 9.4 Too long a wait without an increase in challenge.

So, if we manage to get our students into flow, giving them extensions synchronously will not work. We have to work asynchronously—we have to get the timing right for each student. If students are working individually in their seats, this is almost impossible. Fortunately, thinking classrooms have two things that make this easier. First, students are working in random groups. This reduces what we need to manage asynchronously from 20–30 students to 7–10 groups. Second, students work on vertical non-permanent surfaces. This makes it easier to observe where groups are in their thinking and makes it easier for us to get the timing right.

Aside from maintaining the balance between ability and challenge, we also have to remember that Csíkszentmihályi found two other environmental conditions for the optimal experience to occur—clear goals every step of the way and immediate feedback on action. When we combine these three conditions together, we can begin to see why the jellybeans (Chapter 4), tax collector (Chapter 6), and birthday cake (Chapter 8) tasks worked so well with our students. Each of these tasks has a clear goal, and each of these tasks offers the possibility of immediate feedback on action. In fact, for the jellybeans and birthday cake tasks, the feedback is provided by the task itself—the solution either worked or it didn't. Finally, each of these tasks offers us the ability to increase the challenge as students complete a level.

These same principles can be applied to the idea of curricular thinking tasks introduced in Chapter 1. Let's look at some examples from different grade levels.

First, consider a task used with high school students that began with the following script:

Teacher Let's start with a bit of review. How would I expand $(x + 2)(x + 3)$?

[Teacher writes on the board $(x + 2)(x + 3) =$]

Students $x^2 + 5x + 6$.

[Teacher writes on the board $(x + 2)(x + 3)$
$= x^2 + 5x + 6$]

Teacher Ok. So what if my answer was $x^2 + 7x + 6$? What would the question be?

[Teacher writes on the board $(\ \ )(\ \ ) = x^2 + 7x + 6.$]

$$(x+2)(x+3) = x^2 + 5x + 6$$
$$(\quad)(\quad) = x^2 + 7x + 6$$

This task already has a clear goal—figure out what the two binomials are that, when multiplied, will produce the desired trinomial. Also, by the fact that students can use distribution to check their answer, this task has the ability to provide immediate feedback on actions. All that is missing are the extensions—the ability to increase the challenge as a group's ability increases. Consider the following sequence of such extensions:

1. $(x+2)(x+3) = x^2 + 5x + 6$
2. $(\quad)(\quad) = x^2 + 7x + 6$
3. $(\quad)(\quad) = x^2 + 7x + 12$
4. $(\quad)(\quad) = x^2 + 14x + 24$
5. $(\quad)(\quad) = x^2 + 10x - 24$
6. $(\quad)(\quad) = x^2 + 4x - 12$
7. $(\quad)(\quad) = x^2 - x - 12$
8. $(\quad)(\quad) = x^2 - 2x - 24$
9. $(\quad)(\quad) = x^2 - 6x - 16$
10. $(\quad)(\quad) = x^2 - 0x - 16$
11. $(\quad)(\quad) = x^2 - 25$
12. $(\quad)(\quad) = x^2 - 49$
13. $(\quad)(\quad) = x^2 - 10x + 24$
14. $(\quad)(\quad) = x^2 - 13x + 12$

This sequence was created using two main principles of *variation theory* (Marton & Tsui, 2004). The first principle is that we can only see variation against a backdrop of non-variation. That is, that before something changes, it has to stay the same. We see this in the transition from Task 4 to Task 5. Prior to making the third coefficient negative, we kept it positive for four tasks. The second principle is that only one thing is varied at a time. So, although Task 14 is far from Task 1, at every stage only one thing was varied. First, we varied the number of factors that the third coefficient provided. Then we made the third coefficient negative; then the second coefficient became negative. And so on. Although inspired by variation theory, this is very similar to the idea of number strings (Fosnot & Dolk, 2002).

When we have used this sequence, or one similar to it, with Grade 10 students, we get through the entire sequence in 40–60 minutes. That is, depending on the curriculum, we are able to cover the entire factoring quadratics unit in one lesson. How is this possible? The short answer is that when students are not thinking, everything we teach them is difficult. When students are thinking, however, almost anything is possible. When students are thinking, they are learning and understanding—and this transfers to success.

> Once your students are thinking—both individually and collaboratively—a sequence such as this, used asynchronously to maintain the balance between ability and challenge, allows you to cover a huge amount of content in a single lesson.

This is where you start to earn back the time you spent doing non-curricular thinking tasks. To be clear, this is not something that can be done on Day 1 of starting to build thinking classrooms. But once your students are thinking—both individually and collaboratively—a sequence such as this, used asynchronously to maintain the balance between ability and challenge, allows you to cover a huge amount of content in a single lesson.

Let's look at another script and sequence of tasks that can be used to teach one- and two-step solving of algebraic equations at the middle school level.

Teacher	We are going to play a game of guess what's in my head. I'm going to think of a number, and you are going to guess what it is. To help you make the guess, I will give you exactly one hint.
Teacher	OK—I have my number. Here is your hint—if I add three to my number the answer is 12. Thumbs up if you know my number.

Students	[Students put up their thumbs.]
Teacher	[When enough thumbs are up, the teacher calls on the students.] OK—what is my number?
Students	9.
Teacher	Great. OK—I have a new number. Here is your hint—if I double it and add three my answer is 15. Thumbs up if you know my number.
Students	[Students put up their thumbs.]
Teacher	[When enough thumbs are up, the teacher calls on the students.] OK—what is my number?
Students	6.
Teacher	OK. Before I give you the next one, we have to learn how to write what I just said. [Teacher writes on the board: $\square \times 2 + 3 = 15$.]
Teacher	And before I give you the next one, there are three rules to this task.

1. You can use a calculator.

2. If you use a calculator, you must write down on the whiteboard what you type into the calculator.

3. You have to check your answer by putting it back into the calculator.

Teacher	Here is your next one: $\square + 3.014 = 7.22$. [Teacher randomizes the groups and sends them off to work.]

1. $\square + 3.014 = 7.22$

2. $\square - 15.1 = 7.88$

3. $\square \times 4.25 = 24.8$

4. $\square \div 1.356 = 4.02$

5. $\square \times 2.5 + 3.67 = 18.3$

6. and so on

This sequence, like those before it, adheres to the two principles of variation theory—variation can only be seen against a backdrop of non-variation, and only one element should be varied at a time. We have used this exact script and a similar sequence to move through all of solving one- and two-step equations in a single lesson. In fact,

I have personally taken a thinking Grade 5 class through this script and sequence in 35 minutes, by the end of which every group was solving tasks as complex as $\square \div 15.3 - 8.27 = 3.01$. In our research we have been able to recreate these results hundreds of times. And every time we move through huge amounts of content in a single lesson. This is not to say that every group gets equally far, but every group gets to the point where they are solving two-step equations, or factoring quadratics where the leading coefficient is greater than one. The groups that get there sooner are given increasingly harder and harder questions beyond this to keep them in flow while the rest of the groups catch up.

Now, consider a sequence, adapted from the unusual baker task (NCTM, 2012), that can be used in an upper elementary classroom on the topic of fractions. The premise is that the baker cuts their cakes in different ways every day (see Figure 9.5). What fraction of the cake is each piece?

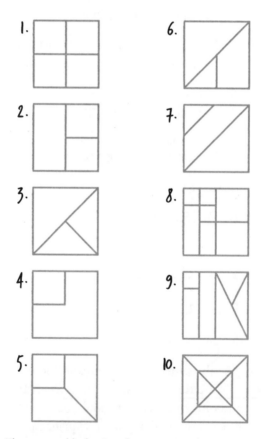

Figure 9.5 The unusual baker's cakes.

For a primary sequence, consider the following tasks, where students are asked to find the next three terms of a number pattern:

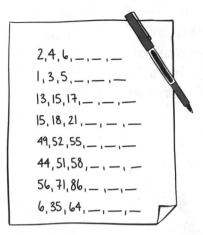

$$2, 4, 6, \_, \_, \_$$
$$1, 3, 5, \_, \_, \_$$
$$13, 15, 17, \_, \_, \_$$
$$15, 18, 21, \_, \_, \_$$
$$49, 52, 55, \_, \_, \_$$
$$44, 51, 58, \_, \_, \_$$
$$56, 71, 86, \_, \_, \_$$
$$6, 35, 64, \_, \_, \_$$

Each of the sequences of curriculum tasks has another quality that is common among them—the increase in challenge from one task to the next is incrementally small. In other words, we are just pushing the envelope a little bit in each step, which helps keep students in flow and reduces the chance of pushing them into frustration by making cognitive leaps too challenging. I refer to such sequences as *thin slicing* (see Figure 9.6). Thin slicing sequences stand in contrast to the *thick slicing* sequence we see in thinking tasks such as tax collector (Chapter 6), ice cream cones (Chapter 5) and wine chest (Chapter 5), where the increase in challenge from one part of the task to the next is much greater (see Figure 9.2).

> Each of the sequences of these curriculum tasks has another quality that is common among them— the increase in challenge from one task to the next is incrementally small.

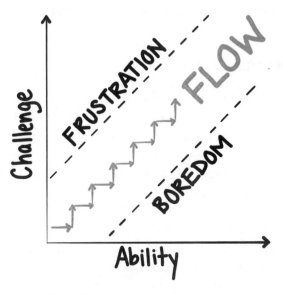

Figure 9.6 Thin slicing sequence of tasks.

> **The benefit of thin slicing sequences of thinking tasks is that it helps avoid frustration.**

As mentioned, the benefit of thin slicing sequences of thinking tasks is that it helps avoid frustration. The drawback is that groups can move through the tasks very quickly, putting an extreme pressure on you to get to groups that are done before they get bored (see Figure 9.4). In the algebra sequence seen earlier, I have seen groups dispense with one task every three to four minutes. If there are 10 groups in the room, getting to each group in time becomes impossible, and groups will start getting bored—unless you have mobilized the knowledge in the room (Chapter 8). Once that happens, students begin to use the autonomy afforded them to keep themselves in flow by stealing the next tasks from groups around them—essentially increasing the challenge of the tasks on their own. You can help make this easier by setting a rule that whatever task they are currently working on needs to be written at the top of whatever vertical surface they are working on—thereby making it easier for others to steal.

When I ran the algebra sequence with that group of fifth graders, I gave out each task to only one or two groups. The rest of the groups got their next task by stealing it from others—when they were ready for it and to keep themselves in flow. This frees the teacher up to spend more time attending to the groups for whom, despite your best efforts, there is an imbalance between their ability and the challenge of the task, and they start to head for frustration (see Figure 9.3). This will happen. And when it does, we need to intervene by giving a hint.

Using Hints to Maintain Flow

> **There are two types of hints—hints that decrease challenge and hints that increase ability.**

It turns out that there are two types of hints—hints that decrease challenge and hints that increase ability (see Figure 9.7). The first of these is quicker to give, and either requires you to give a partial answer to the question students are working on or shift them to an easier task. The second type of hint—increase ability—takes longer and requires you to either remind them of a strategy or give them a strategy. Other than how long it takes to give these hints, the main difference is that hints that decrease challenge are only useful in that moment, whereas a hint that increases ability continues to be useful even as students move on to the next task.

For example, if the factoring quadratic sequence is used, you will eventually get some groups working on $6x^2 + 13x + 5$. If a group is struggling with this one, you can tell them that the first term in the binomials will be $2x$ and $3x$, or write $(2x +)(3x +)$ on the board for them. This is an example of a hint that reduces challenge. Alternatively, you can ask them to tell you how they think about the last term in the trinomial (5). When they tell you that that term is the product of the second terms in the binomials, you can either smile and walk away, or you can mention that maybe they should use that same thinking to consider the leading coefficient in the trinomial. This hint takes longer, but it mobilizes and repurposes knowledge they already have—it is an example of a hint that increases ability.

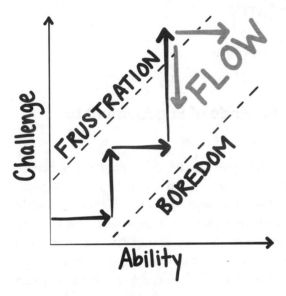

Figure 9.7 Two types of hints.

Obviously, hints that increase ability are better in the long run. But frustration is not about the long run. Frustration is an intensely negative emotion that needs a rapid intervention, and sometimes the best way to intervene quickly is to reduce the challenge. Of course, the imbalance can be tipped in the other direction as well—the ability of the group exceeds the challenge of the tasks, and they start to head for boredom. The obvious thing to do in these situations is to increase the challenge of the task (see Figure 9.8). This is, in essence, the same as using extensions to maintain flow (see previous section).

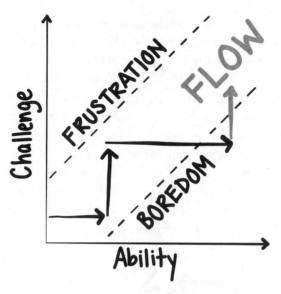

Figure 9.8 Increase the challenge.

Using Shifts in the Mode of Engagement to Maintain Flow

> Rather than shift the task the group is working on, we shift their *mode of engagement* with the same task.

There is one more way to regain the balance between the ability of the group and the challenge of the task at hand. Rather than shift the task the group is working on, we shift their *mode of engagement* with the same task (see Figure 9.9). For example, when students are solving a task, their mode of engagement is *doing*—they are *doing* the task. This is the easiest way to engage with a task. If I tell a group that has finished that they are wrong, or ask them if they would bet $100 on their answer and then walk away, I just shifted their mode of engagement from doing to *justifying*. Justifying is more challenging and involves students convincing themselves that they are correct. When they are convinced and call me back to tell me they are correct, I may ask them to *explain* to me how they know they are correct. Explaining is harder than justifying, as it requires the articulation of thought for an audience outside of those who did the original thinking. The group will likely not be good at explaining the first time around, so I may tell them to come get me when they can explain to me in a way that I understand. When this happens and they do a good job explaining, I may point them toward another group and ask them to help that group by *teaching* them something. If we subscribe to the notion that teaching is different than telling or explaining, then this is another increase in challenge. Once the group has finished teaching, I may ask them to *create* a new task for that group. Creating is the most difficult mode of engagement, as it requires the group to not only see the didactics of the situation, but also the pedagogical needs and affordances of the next task.

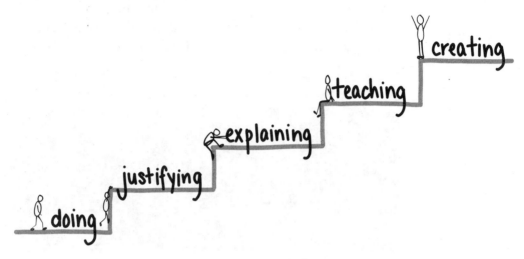

Figure 9.9 Modes of engagement that increase challenge.

My research showed that this sequence of modes of engagement—*doing, justifying, explaining, teaching, creating*—not only increases challenge, but does so in a way that can continue to engage, or reengage, even the strongest group. It also explains why students often have a difficult time explaining their thinking—we are asking them to go straight from doing to explaining. We need to first ask them to justify their thinking.

The bottom line with all of this is that the goal of building thinking classrooms is not to find engaging tasks for students to think about. The goal of thinking classrooms is to build engaged students that are willing to think about any task. We achieve this through using thinking tasks (Chapter 1), visibly random groups (Chapter 2), and vertical non-permanent surfaces (Chapter 3); defronting the room (Chapter 4); answering only keep-thinking questions (Chapter 5); giving tasks verbally, early, with the students clustered around (Chapter 6); giving check-your-understanding questions (Chapter 7); and mobilizing the knowledge in the room (Chapter 8). Once we have this, we can point the thinking classroom

> The goal of building thinking classrooms is not to find engaging tasks for students to think about. The goal of thinking classrooms is to build engaged students that are willing to think about any task.

at curriculum. Because curriculum tasks are not innately engaging for students, we need to manufacture engagement through giving clear goals, ensuring there is an ability to get immediate feedback on actions, and asynchronously maintaining the balance between the group's ability and the challenge of the task at hand through the use of hints and extensions (see Figure 9.10). When we do this, not only do we make curriculum tasks more engaging, but we also create the possibility of covering a large amount of curriculum in a short amount of time.

Figure 9.10   Students in flow on curricular tasks.

Q If this is so effective, should I be doing this every day?

A Yes and no. No because you cannot start the school year like this. You need to begin by using highly engaging non-curricular thinking tasks to build the culture of thinking in your classroom. Non-curricular tasks are also good to use whenever you introduce a new thinking practice into your classroom—this is why every previous chapter ends with one. For the rest of the time, the answer is yes. If you want your students to be thinking they need to be engaged. And for them to be engaged they need to be in flow. This chapter outlines the best way we have found to keep students in flow while working on curricular content.

Q In all the sample examples that were provided in this chapter, the first tasks students were asked to solve was very easy. Doesn't this risk them becoming bored?

A If the students are relying on you to get to them before they can go on to a more challenging task, then yes. But if they are stealing tasks from others, then this is not an issue. On the positive side, starting with a very easy task increases the chances that all groups will be able to start.

Q A lot of topics we teach draw on students' prior knowledge. But sometimes we teach topics that are brand new to the students. Normally this requires a long lecture to orient them toward the topic and teach them how to do the tasks that are coming. How do we deal with these situations inside of the thinking classroom framework?

A Prior to reading this chapter, many teachers would say exactly the same thing about both solving one- and two-step equations and factoring quadratics. Yet, neither of these needed a long introduction to teach students how to do it. As teachers, we often design our lessons to prepare students to answer the hardest task they will face. When using flow, our teaching needs only to prepare them to answer the first task they will face and then count on the fact that they will learn something during that task that will help them with the next task. This does not mean that you can't just tell them something to get them started, but based on the results from Chapter 6, you have a maximum of five minutes to do it. In the Chapter 1 FAQ, I gave an example of a script for introducing the Pythagorean Theorem in less than five minutes.

Also, do not forget that while students are working their way through the sequence of tasks, you are still in the room. The instruction you would normally give at the beginning of the lesson can now become hints that you use when needed to keep groups moving further up the flow channel.

Q Knowledge mobility aside, wouldn't it just be easier to give each group the list of tasks they need to work through?

A We thought so too at first. But repeated attempts at making this work always produced the same result—a dramatic shift from trying to learn from the solving of the tasks toward *just get it done*. That is, the students stopped caring about being in the moment and making sure that everyone in the group understood. Instead, the list provided a finish line, and they wanted to get across it as fast as possible. So, usually the strongest member of the group just took the marker, and off they went.

Incidentally, we saw the same results when we allowed groups to sit down when they had finished our sequence of tasks. The fact that groups got to be finished indicated there was a finish line, and the race was on. In addition, groups who did not finish began to feel anxious about their work and the fact that they were still standing when others had gotten to sit down.

Q One of the elements needed in order to make flow work is immediate feedback on action. The factoring and algebra examples you gave have this built in, but what do I do when the task does not provide that sort of feedback?

A The task is not the only, or even best, source of feedback available to students. The best sources of feedback come from within the group and from the groups around them. Multiple minds thinking together about the same task often provide all of the feedback that is needed as they check their own work. Another source of feedback in the room is you, as the teacher. However, there is a delicate line between providing feedback and answering stop-thinking questions. As mentioned in Chapter 5, an easy way to avoid falling onto the wrong side of that line is to enter a group with questions rather than answers—"Why did you do that?" "Can you tell me what you are doing here?"

Q The explanation of flow is very clear—for an individual. But a group is not a single entity. It is made up of three individuals, all of whom are unique. How can we guarantee that they are all in the same place on this graph?

> A group is more than a collection of individuals—it is a collective. And when a collective is working on something new, they often merge into one entity.

A Students are all unique and, as a result, there is a strong chance that each student is in a different place on the graph. But a group is more than a collection of individuals—it is a collective. And when a collective is working on something new, they often merge into one entity. Take the example of factoring quadratics. This is a new topic to students. When we watch groups work on this, we often observe students of varying abilities coming together in a group and performing as one—completing each other's thoughts and moving forward with synergy.

Of course, this does not always happen, and there may be a student who is falling behind the rest of the group. Sometimes it is two students falling behind a single student who has already built understanding and is wanting to storm ahead. In these circumstances you simply give the marker to the student who is falling behind and make it clear to the stronger student that the group does not get to move to the next task until everyone in the group understands what is happening on the current task. Careful questioning on your part can verify whether this is true.

If there is a student who is operating well above the rest of the group and wants to teach by telling, then have a private conversation with that individual about the different modes of engagement and how you are challenging them to *teach* rather than *do*. Alternatively, if there is a student operating well behind the rest of the class, then working with that student in the period before to prepare them for the coming task is helpful in making them a relevant member of the group. We call this *preloading*.

Q What if a group struggles on one of the tasks in the sequence, but eventually answers it? Do I still give them the next task?

A No, if a group is struggling, then the next task they get is similar to the one they just completed. The completion of a task does not necessarily mean that their ability has grown as much as it can. Another task at the same challenge level will allow their ability to grow more before the challenge increases. What this means is that you need to enter the lesson not just with a single sequence of progressively more challenging questions, but also a sequence of parallel tasks of comparable challenge level.

Q I have noticed that sometimes a group is working on a task that is too hard or too easy for them, but they still manage to stay in flow. What is that about?

A It turns out that Figure 9.1 is incomplete. My research on building thinking classrooms revealed that there are actually two more regions in the flow diagram—*perseverance* and *patience* (Liljedahl, 2018). The regions act as buffers between flow and frustration and flow and boredom (see Figure 9.11) and allow groups to operate out of balance without disengaging from the task. This gives groups time to autonomously pull themselves back into flow by looking to the groups around them for a hint or an extension.

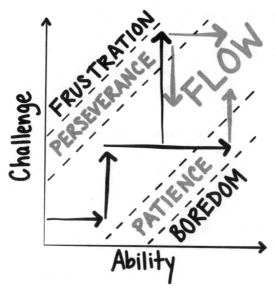

Figure 9.11 Perseverance and patience.

Q I feel like my students currently don't have any perseverance or patience. How do I increase this?

A Both of these increase from repeated opportunities to work in flow. This is another reason why it is so important to use highly engaging non-curricular thinking tasks when you begin to build a thinking classroom—and occasionally thereafter.

Q You mentioned how important it is to have a carefully constructed sequence of tasks. At the same time, you mentioned that we should rely on students' autonomy to steal the next task. What if students pick the tasks out of sequence? Doesn't that mess up their flow?

A In the beginning it is very important that the sequence of tasks is maintained. But as students' perseverance and patience increases, the sequence becomes less critical as the groups become more able to cope with imbalances between their ability and the challenge of the task at hand.

Q Where can I get such sequence of tasks with progressively increasing challenge along with parallel tasks?

A The lists of exercises in textbooks are, for the most part, designed in exactly this way. But be critical. Textbook exercises are designed for practice and not thinking and learning. So, they do not always adhere to the ideas of variation discussed earlier. But, don't underestimate your own ability to make such a sequence. Teachers who are teaching a curriculum that they are familiar with (see Figure 9.12) can sit down and create these sequences for themselves without much effort. Regardless, go into the lesson with a longer sequence than you think you will need. It is better to not complete a sequence than it is to have your students shoot through your planning in the first six minutes—which has happened many times.

Teachers who are teaching a curriculum that they are familiar with can sit down and create these sequences for themselves without much effort.

UNIT 8 DERIVATIVES

$$y = e^{g(x)} \longrightarrow y' = e^{g(x)} \cdot g'(x)$$

$$y = \ln g(x) \longrightarrow y' = \frac{1}{g(x)} \cdot g'(x)$$

$$y = b^{g(x)} \longrightarrow y' = b^{g(x)} \cdot g'(x) \cdot \ln b$$

$$y = \log_b g(x) \longrightarrow y' = \frac{1}{g(x) \cdot \ln b} \cdot g'(x)$$

$$y = \sin[g(x)] \longrightarrow y' = \cos[g(x)] \cdot g'(x)$$

$$y = \cos[g(x)] \longrightarrow y' = -\sin[g(x)] \cdot g'(x)$$

$$y = \tan[g(x)] \longrightarrow y' = \sec^2[g(x)] \cdot g'(x)$$

$$y = \cot[g(x)] \longrightarrow y' = -\csc^2[g(x)] \cdot g'(x)$$

$$y = \sec[g(x)] \longrightarrow y' = \sec[g(x)] \tan[g(x)] \cdot g'(x)$$

$$y = \csc[g(x)] \longrightarrow y' = -\csc[g(x)] \cot[g(x)] \cdot g'(x)$$

Figure 9.12 Sequence of curricular concepts that can be used to build a flow sequence of tasks in calculus.

Q You began this chapter by talking about differentiation. Differentiation is normally seen as providing students with different tasks and activities based on their abilities. Does your use of flow put forward a different way to operationalize differentiation?

As in all forms of differentiation, the teacher plays a vital role in this process, deciding when a group needs a hint or an extension, or to maybe do another task of the same challenge.

A Yes. In this framework every group starts in the same place—with the same task. What is differentiated, then, becomes more about the timing and pacing that each group moves through the sequence of tasks. However, as in all forms of differentiation, the teacher plays a vital role in this process, deciding when a group needs a hint or an extension, or to maybe do another task of the same challenge. This differs from the more conventional idea of differentiation in that it is based on the in-the-moment information about how a group, and the individuals in that group, are performing rather than the preconsidered anticipations, expectations, and sometimes even assumptions or biases about how they may perform.

SUMMARY

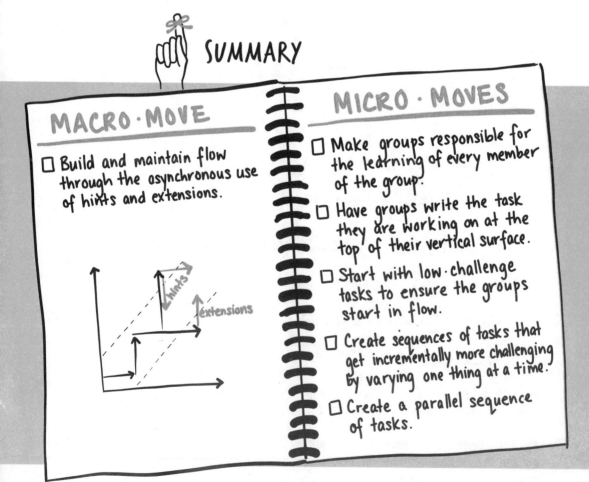

MACRO·MOVE

☐ Build and maintain flow through the asynchronous use of hints and extensions.

MICRO·MOVES

☐ Make groups responsible for the learning of every member of the group.

☐ Have groups write the task they are working on at the top of their vertical surface.

☐ Start with low·challenge tasks to ensure the groups start in flow.

☐ Create sequences of tasks that get incrementally more challenging by varying one thing at a time.

☐ Create a parallel sequence of tasks.

QUESTIONS TO THINK ABOUT

1. What are some of the things in this chapter that immediately feel correct?

2. Pick a topic that you have just finished teaching, and try to build a sequence of incrementally more challenging tasks that cuts through a part, or all, of the topic.

3. Now do the same thing for the next topic you will be teaching.

4. Think about a topic that you believe is brand new to students. What is the minimum set of instructions that you need to give in order to prepare students to be able to think their way through the first task you would ask them to do? What can students learn from this first task?

5. Do you think your students have developed the autonomy they need to allow them to help themselves to stay in flow? If not, reread Chapter 8, and think about how you can continue to further their growth in this area. This does not mean you can't start to play with creating and maintaining flow. Flow is a great context in which to keep working on mobilizing knowledge to build students' autonomy.

6. Have you seen instances where your groups exhibit tremendous perseverance or patience while working in a thinking classroom setting? If so, could they have done this at the beginning of the school year?

7. What are some of the challenges you anticipate you will experience in implementing the strategies suggested in this chapter? What are some of the ways to overcome these?

 TRY THIS

The following sequences of highly engaging non-curricular thinking tasks have been shown to easily put students into, and keep them in, flow.

Grades K–3: The Answers Are

Using each of the numbers from 1 to 10 exactly once and each of the operations + and − at least twice (one will be used three times), make five expressions whose answers are 17, 17, 8, 1, 2. An expression, in this case, is defined as two numbers and an operation. A possible script for introducing this task is:

Teacher	Today we are going to build some expressions. Each expression is made up of two numbers from this list [teacher points at list of numbers 1, 2, 3, 4, 5, 6, 7, 8, 9, 10] and one of these operations. [Teacher points at list of operations + + − −.] Can someone please tell me an expression?
Student	8 + 1.
Teacher	OK. The answer for this is 9. [Teacher writes 8 + 1 = 9.] I forgot to mention that now the 8 and the 1 and one of the + is now gone. [Teacher crosses these out on the board.] Can someone give me another expression?
Student	10 − 1.
	[. . . and so on]
Teacher	Ok. So now we have run out of operations, but we still have two more numbers [Teacher points at the 3 and the 2.] So, let's make one more expression, and you can use one of the operations + or − a third time.
Student	3 − 2.
Teacher	OK. [Teacher writes 3 − 2 = 1.] We now have five expressions [teacher points at the five expressions] and five answers [teacher points at the five answers]. And these answers came from following two rules. We had to use every number from 1 to 10 exactly once, and we had to use addition and subtraction each at least twice. And if we follow these rules, we will get five answers. So, if we know these rules and all we had were these answers [teacher erases the expressions leaving just the answers], could we figure out what the expressions were? And, of course, these are not the answers I care about. [Teacher erases

the answers.] These are the answers I care about.
[Teacher writes 17, 17, 8, 1, 2 on the board.]

[Teacher makes random groups]

The sequence of answers that follows is

13, 9, 13, 1, 13

13, 1, 9, 1, 17

15, 1, 1, 1, 19

2, 2, 7, 7, 7

2, 2, 2, 3, 8

3, 3, 3, 3, 19

Grades 4–12: The Answers Are

Using each of the numbers from 1 to 10 exactly once and each of the operations +, −, ×, and ÷ at least once (one will be used twice), make five expressions whose answers are 5, 8, 13, 24, 20. An expression, in this case, is defined as two numbers and an operation. The sequence of answers that follows is

17, 2, 21, 3, 2

10, 14, 1, 20, 16

3, 3, 3, 3, 24

2, 2, 2, 2, 9

2, 3, 7, 7, 7

1, 2, 3, 4, 5

A possible script for introducing this task can be made by modifying the script above.

CHAPTER 10

HOW WE CONSOLIDATE A LESSON
IN A THINKING CLASSROOM

Consolidation is an important part of every lesson. Through consolidation we are able to bring together the disparate parts of a lesson and help students to reify their experiences into a cohesive conceptual whole. But what does this look like in a thinking classroom, where the effort to keep groups in flow is more important than keeping all the groups together, and where student autonomy allows groups to solve tasks in very different ways? In this chapter you will learn what consolidation needs to look like in a thinking classroom, not only from the perspective of pulling a wide variety of student work together, but also how to do so while keeping thinking as a central focus.

To reify: is to make concrete and real something that is abstract or intangible.

THE ISSUE

When I spent time visiting those 40 classrooms at the outset of my research, I saw a lot of consolidation. Typically, it occurred after a now-you-try-one task. Sometimes this consolidation would involve students sharing their solution, but more often than not it involved the teacher going through the solution, step-by-step on the board. In the introduction, you learned how this rhythm of giving a now-you-try-one task, waiting (4 minutes and 22 seconds), and then going over it enabled- nay, encouraged— students to stall or fake in anticipation of not only getting the answer, but getting the *best* answer from the teacher.

For the most part, consolidation after a now-you-try-one task consisted of what Alan Schoenfeld calls leveling to the top (1985). What this means is that, irrespective of where students are in their thinking or their solution process, the teacher goes over the most advanced and most nuanced aspects of the solution. They do so in the belief that in order to be able to do the next now-you-try-one task, students need to know how to do this one, and if they have not gotten it after 4:22, we will just give it to them so they are ready to move on to the next task. In essence, we level to the top in an effort to lift everyone up to the top.

THE PROBLEM

The problem is that it doesn't work. Unless a student is close to the answer in their thinking, then a leveling to the top is too big a cognitive jump for them to take. And the result is the exact opposite of what we were hoping for—rather than preparing the students for the next task, they are actually less

> If all students could learn by having us just tell them how to do it, we would not have any of the problems that we have in mathematics education today.

prepared for it and are even less likely to be able to solve it. If all students could learn by having us just tell them how to do it, we would not have any of the problems that we have in mathematics education today. For over one hundred years the dominant pedagogy was teaching through telling. If that had worked, then all students would have been in our highest streams, and all students would have gotten the highest marks. But that has not been the result. Student difficulty with mathematics has been a pervasive and systemic problem since the advent of public education—not because students can't learn mathematics, but because, by and large, students can't learn by being told how to do it.

> Students began to mistake being shown how to do it for learning, and they mistook having it in their notes for knowledge.

More specifically related to thinking, I also noted that when teachers leveled to the top, this became a non-thinking activity both for the students who had already gotten there and for the students who were not even close. For this latter group, rather than thinking, this simply became a mindless note-taking exercise. I'll discuss this more in Chapter 11, but for now I can say that students began to mistake being shown how to do it for learning, and they mistook having it in their notes for knowledge. This came through in classroom observations and student interviews over and over again. If you have been implementing the thinking classroom in concert with reading this book, it has likely come through in your experiences as well. When students ask, "When are you just going to go back to teaching us math?" what they are really asking is when are you going back to telling (showing) us how to do it so we can just write it down in our notes.

TOWARD A THINKING CLASSROOM

So, if leveling to the top does not work, what would be the effect of leveling to the bottom? And what, exactly, would that look like? This is exactly what we began experimenting with. Our starting point came from the idea that if leveling to the top is about presenting a solution of where we want all students to get to, then consolidating from the bottom must start with the presentation of solutions that all students got to. That is, the consolidation follows the same path as the extensions of increasingly challenging tasks that were used to create and maintain flow while the students worked through the task(s).

If this were a non-curricular task such as the gold chain task (see Chapter 4), then we might start by discussing what a solution looks like if we cut every second link. This is a point that every group starts at, and it turns out to not lead to a viable solution. From there we may discuss the

> **Consolidating from the bottom must start with the presentation of solutions that all students got to.**

solution where we cut every third link and use the intact two-link segments to pay for both the room and the cut on the days that a cut is made. With your use of hints, every group would have gotten to this solution. This is followed by a discussion of a solution where we start getting change for our gold links. So, for example, if on a certain day we owe two links of gold, we pay with a length of four links and get two single links as change. So, rather than starting the consolidation with the change model—the model we want all students to get to—we started the consolidation with the solution that every group did first.

If it is a set of curricular tasks, such as the factoring quadratics sequence of tasks (Chapter 9), then the consolidation would start with how to factor quadratics where all the coefficients are positive and the leading coefficient is one—something all groups would have been able to do. From here we would discuss how to factor quadratics where the last coefficient is negative, and so on. For the unusual baker sequence of tasks (Chapter 9), you would begin by reviewing the first two tasks (see Figure 10.1). This will allow you to reemphasize the importance of the number of pieces as well as the relative sizes of the pieces when trying to determine what fraction of the whole cake each piece is. From there you would jump to the fourth and fifth cakes, for which you would review the addition of extra lines for the purpose of creating equal-size pieces, and so on.

When we began to experiment with this consolidation technique, we immediately noticed that rather than disengaging during consolidation, all students were with us from the beginning, and more students stayed engaged. For the students who stayed engaged, consolidation stopped being an extension of the lecture where worked examples are demonstrated. Instead, it started becoming a reifying activity where their ideas were valued and expanded on. In this way, more thinking, and thus more learning, was occurring. Things were sticking better.

Figure 10.1   The unusual baker's cakes.

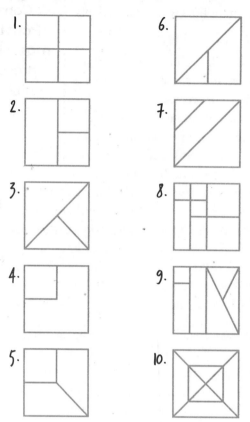

There are three ways in which consolidation from the bottom can take place:

1. The teacher leads a general discussion about the task(s) and solution(s) but writes nothing down.

2. The teacher leads a detailed discussion of the task(s) and solution(s) while recording on the board what is being discussed.

Source: Photo courtesy of Judy Larsen. Used with permission.

3. The teacher leads a detailed discussion of the task(s) and solution(s) using student work on the vertical surfaces to work through the different layers of the solution.

The first of these methods is most useful when talking about big ideas and general strategies that have emerged out of the student activity. For example, the tax collector task (Chapter 6) can be consolidated in this way as the class discusses the strategies for choosing the first and second envelopes. The second method is suitable when more detail is required—such as when consolidating ideas around how to add two-digit numbers or when completing the square. To be clear, both the first and second methods are not lectures where the teacher does most of the talking. They are discussions where the teacher asks very focused questions and the students contribute ideas—ideas that are reified and formalized through the teacher's reframing of things that are being offered by the students. The difference between the two is that the teacher records these ideas in the second method, whereas in the first, the ideas exist only as a discussion. The third method, which turned out to be the most effective method for maintaining engagement, is used in the same circumstances as the second method, but rather than having the teacher write on the board, the existing work of students is used to demonstrate the details.

Regardless of the method teachers used, they helped us find some refinements that drastically increased student thinking during the consolidation process. For example, we found that having the students standing in a loose cluster around the teacher significantly increased attention and engagement during the consolidation process. In one instance we repeated the study discussed in Chapter 6, where we simply documented students looking at their cell phones during consolidation. In the lesson where the students sat during consolidation, almost 50% of the students looked at their cell phones for more than 50% of the time. When these same students were standing during the consolidation, only one student looked at their cell phone, and then only for 30 seconds. It turns out that having students stand generates more engagement both when being introduced to the task (Chapter 6) and when debriefing the task.

Another refinement that made a big difference in student engagement has to do with how much time students spend at each level of the task or task sequence. When leveling to the top, 100% of the time is spent on the solution or method that the teacher wants the students to ultimately reach. When consolidating from the bottom, there is a gradual movement through each level of the task, with more time spent at the first level, and then a decreasing amount of time spent on each successive level until the highest level is just barely touched on. In essence, as you move through the different challenge levels of the flow diagram, you spend less and less time focusing on the processes at each level (see Figure 10.2).

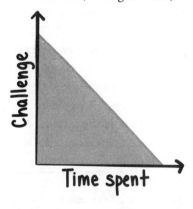

Figure 10.2   Time spent consolidating as challenge increases.

What this does is allow more students to stay with the discussion longer, as the teacher is going over aspects of the tasks that more students were able to do. For example, consider the following flow sequence for adding two-digit numbers:

1. $10 + 10 =$

2. $11 + 10 =$

3. $14 + 10 =$

4. $14 + 20 =$

5. $17 + 20 =$

6. $33 + 20 =$

7. $24 + 50 =$

8. $24 + 51 =$

9. $24 + 52 =$

10. $24 + 57 =$

11. . . .

Spending time discussing and reifying some of the foundational aspects of adding 10s that emerge out of the first three questions allows every student to participate in the discussion. As the tasks get more nuanced and fewer students have been able to complete them, you spend less time focusing on them, until you get to the most advanced tasks, when you may say something like this:

> Teacher   And this idea of adding 10s doesn't change, no matter what the 1s are. But, when the 1s add to more than 10, we will have to rethink how many 10s there are.

As mentioned, consolidating through the use of student work (Method 3) was the most effective at creating and maintaining student engagement. This method is often referred to as a *gallery walk*. In a gallery walk, rather than the work to be discussed coming to the students as in the first two methods, here the students literally walk to the work to be discussed. With this method, too, we found refinements that increased both engagement and the amount of time students spent thinking. The first and most impactful of these refinements was the realization that when students spoke to the whole class about their own solutions, very few, if any, of the other students listened. This was such a shocking result that we sought to verify it in contexts other than consolidation—even contexts outside of the thinking classroom research. What we found was that during whole-class discussion, unless there is a punitive structure in place forcing students to listen attentively to other students' presentations of their solutions, very few students did.

Our fix was that when discussing student work, we ask other members of the class to try to explain what the group was thinking whose work we were discussing.

Teacher Can someone NOT in this group tell me what this group was doing here? [Teacher points at a specific part of the board.]

Teacher Talk amongst yourselves to see if you can figure it out.

This approach created a space where students had to try to merge what they were seeing in the written traces of a group's thinking with their own thinking about the same task. Coupling this with opportunities to discuss this mergence in small groups created a space that not only asked them to think, but necessitated thinking. In comparison to simply having the group that produced the work explain the work, this approach changes consolidation from telling to thinking—from passively receiving knowledge to actively thinking about the work at hand. And it positions knowledge as tentative, negotiable, and fallible rather than absolute, definitive, and accurate—a positioning that offers more space for thinking to happen.

> Rather than explaining one's own work, trying to decode someone else's work changes consolidation from telling to thinking.

Another refinement that increased thinking and engagement came from the careful selection and sequencing of what students would attend to during the consolidation. The gallery walk was not a random walk but a focused guided tour through the different levels of the flow sequence. Prior to the consolidation, the teacher not only carefully selected and sequenced the vertical surfaces they would take the students to, but also chose in advance what parts of a group's work they would attend to when visiting that work. This, coupled with the focused questions and reifying language, allowed the teacher to build a cohesive narrative through the consolidation as they took the students through the increasing challenges of the flow sequence of the task or tasks. This refinement also provided a further insight into why having students present their own solutions was ineffective—they would distort the narrative that the teacher was trying to build.

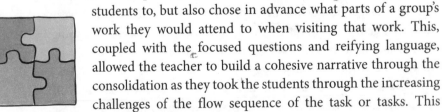

This selecting and sequencing is both similar to and different from Smith and Stein's (2011) notion of selecting and sequencing. They are similar in that they both try to achieve a sensible sequence that moves the learners through various ways of thinking about a task or tasks. The difference is that, whereas Smith and Stein do both the selecting and sequencing in the moment, within a thinking classroom, the sequencing has already been determined within the task creation phase—created to invoke and maintain flow. What is left to do is to select the student work that exemplifies the mathematics at the different stages of this sequence. Another difference is that, whereas Smith and Stein have students present their own work, in the thinking classroom the decoding of students' work is left to the others in the room.

The final refinement that increased student engagement was to keep the students moving. That is, selecting vertical surfaces in a sequence that required students to walk to different parts of the room was more effective for maintaining student engagement. In essence, the gallery walk needed to be a gallery *walk*. The more steps the better.

> The gallery walk needed to be a gallery *walk*.

Given the effectiveness of the guided gallery *walk* in maintaining student engagement and necessitating student thinking, this should be the most frequently used method of consolidation. This is not to say that the other two methods are ineffective, but they should be used only in situations where the guided gallery walk is not suitable.

Q If the guided gallery walk is to be effective, we need to have student work from every task, or every solution strategy, present on the boards at the time of consolidation. How do we ensure this when students have the option of erasing whenever they want?

A Preparation for consolidation begins very early on in the flow sequence. During this time, you, as the teacher, need to be on the lookout for student work that you would like preserved for the consolidation. If you see such work, simply draw a box around it with your red marker and ask the students to not erase it. In the end, you may not use that particular work for the consolidation, but this practice ensures you have the options you need when it comes time to build your consolidation sequence.

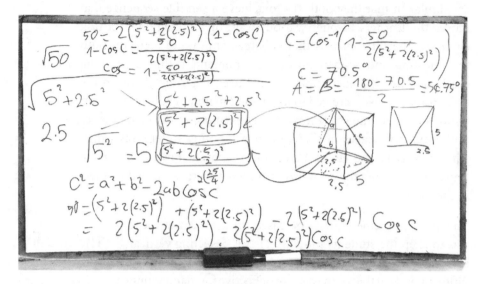

Q Rather than asking students not to erase work, wouldn't it just be easier to take pictures of work and use those for the consolidation?

A Taking pictures is a great idea. The only caution is that pictures create a temptation to use them as a slide show during the consolidation—with students sitting in their seats. Regardless of what method you choose to consolidate, or how you choose to preserve the students' work, putting students back in their seats drastically diminishes the thinking and engagement during the consolidation. Keep them standing.

Q What if a specific approach to a task that I want to discuss during the consolidation is not appearing in any group's work?

A If this happens you have two options—the first of which is to simply discuss (Method 1) or demonstrate (Method 2) this approach as part of the consolidation process. A far more effective option, however, is to drop a hint with one or two groups during the flow sequence that they may want to try this approach. This almost always gets what you want up on the vertical surfaces. What this means is that planning for consolidation begins early on in the flow sequence and requires you not only to lock in student work, but also to subtly plant the ideas you want to appear.

Q What if a group has a great approach to a task and I want to include it in the gallery walk—but their work is incomplete or incoherent?

A There are two things you can do in this situation. First, you could ask the group to clean up their work a bit before the consolidation. Second, you can add notation to their work during the gallery walk as you are facilitating a whole class discussion. It is perfectly OK to add notation with your red marker to any student work during the gallery walk. However, it is not OK to erase students' work. Erasing devalues their work and should only be done when absolutely needed and only with permission.

Q All of the preparation for the guided gallery walk seems daunting. How do I keep track of it all?

A It is daunting at first, but teachers report that once started, they quickly get very good at it. Your red marker is highly visible and allows you to see the traces of your efforts to lock in student work and drop hints. At some point prior to beginning the consolidation, you build your narrative by simply going around and numbering the vertical surfaces in the order you want to discuss them while at the same time drawing boxes around the aspects of the work you want to discuss during the guided gallery walk.

Q With all this focus on student work, shouldn't we be sensitive about their feelings if we are looking at something that is incorrect?

A We should always be aware of students' sensitivities. The fact that groups are not allowed to talk about their own work affords them a certain anonymity. We have also found that pulling all students to the center of the room or some vertical-surface-free neutral corner before the consolidation begins and having a discussion about what they were asked to solve—"Can someone rephrase what you were being asked to do?"—severs their attachment to, and ownership of, the work. Not only does this increase the anonymity of the work, but it opens up the possibility for the teacher to take ownership of the gallery for their own purposes. By first severing the ties between groups and their work, it allows the teacher to lead discussions about work rather than about students.

Q What if I don't know exactly what a group was thinking? Should I still include it in the gallery walk?

A The purpose of the discussion of student work during the gallery walk is not to figure out, with absolute certainty, what a group was thinking. The goal is to use their work, the traces of their thinking, to get the whole class to think and explain. So, not knowing exactly

> The goal is to use students' work, the traces of their thinking, to get the whole class to think and explain.

what a group was thinking is OK. But if you do want to know, you can always ask them prior to the start of the consolidation. You are still active in giving hints and extensions, and part of this requires you to interact with groups—especially groups where you do not understand what they are doing. Preparation for consolidation is added to the important work of creating and keeping students in flow—it does not displace this work.

Q Should I try to select something from every vertical surface for the guided gallery walk?

A No. This would create too much redundancy and would cause the consolidation to take too long. Over time, every student will have their work honored. It doesn't have to happen every day.

Q In this chapter you talk about consolidating and reifying at the level where students got to. And you also talked about leveling to the top as something that doesn't work. Does that mean we can never lift students' understanding above the level they got to?

A We can. But there is a limit to how far we can lift their understanding at a single point in time. What we saw in the research is that when we consolidated from the bottom, moving up through the levels that they had already worked through, we were able to lift their understanding up beyond where they got to. Leveling to the top doesn't work because it starts the consolidation above (often too far above) where the students reached, rather than collecting them at the bottom, reifying ideas and terminology, and then moving upward.

Q If the guided gallery walk is the most effective, when are the other two methods of consolidation appropriate?

A The first method—discussion without notation—is really useful when discussing global strategies that have emerged from the student work—"What should we do first when making a graph?", "When do we need to regroup?" It is also an important part of the reifying discourse that is happening the whole time during a guided gallery walk. The second method—discussion with notation—can also happen at any point within a guided gallery walk as a way to apply a group's strategy to a new task—"So, how would this group add these two numbers?" As a standalone strategy, however, it should be used sparingly, as its similarity to a lecture can easily put students into a passive mode of receiving knowledge rather than an active mode of thinking.

SUMMARY

MACRO · MOVE

☐ Consolidate from the bottom.

#1
$$A = l \times w$$
$$= 7 \times 10$$
$$= 7 \times 2$$
$$= 140$$

140
+ 120
+ 84
344

$$6 \times 10 = 60 \times 2 = 120$$
$$6 \times 7 = 42 \times 2 = 84$$

#2
$$SA = 2(lw + wh + lh)$$
$$= 2(7 \cdot 6 + 6 \cdot 10 + 7 \cdot 10)$$
$$= 2(42 + 60 + 70)$$
$$= 2(172)$$
$$= 344 \, cm^2$$

7 cm
6 cm
10 cm

MICRO · MOVES

☐ Lock in student thinking by drawing a box around it with your red marker.

☐ Use hints to get missing ideas up on the vertical surfaces.

☐ Select and sequence students' work for guided gallery WALK.

☐ Keep the students standing.

☐ Keep the students walking.

☐ Spend more time on the foundation ideas at the beginning of the consolidation.

☐ Do not let students present their own work.

QUESTIONS TO THINK ABOUT

1. What are some of the things in this chapter that immediately feel correct?

2. In this chapter you learned about consolidation as moving through the flow levels of a task or sequence of tasks. And while doing so, to start slow and go faster as you go. This means that the most nuanced and sophisticated solutions will get the least attention. How do you feel about this?

3. In the FAQ, it was mentioned that taking pictures is a very good idea. But you were cautioned against having

the students sit while showing these pictures. So, what are pictures good for, and can you think of ways to use these pictures in a thinking classroom without allowing them to take away opportunities to think?

4. Planning and preparing for consolidation while trying to maintain flow can be daunting. What are some things you can do ahead of time so that it will become less daunting?

5. In Chapter 6 you were presented with results that showed that we need to get the students thinking about a task within the first five minutes of class. This removes from our practice the ability to *teach* at the beginning of the lesson. This chapter, on consolidation, offers us a place where that teaching can now occur. How do you feel about consolidation—at the end of the lesson—as *teaching*?

6. What are some of the challenges you anticipate you will experience in implementing the strategies suggested in this chapter? What are some of the ways to overcome these?

 TRY THIS

The following are tasks that produce a variety of different solutions and solution pathways. As such, they are ideal for practicing consolidating from the bottom.

Grades K–5: Farmer John

A farmer has some chickens and some pigs. One day they notice that their animals have a total of 22 legs. How many chickens and how many pigs might they have? Can you come up with another solution? And another? Can you come up with all the solutions? How do you know that you have all the solutions?

Grades 6–9: Painted Cube

A 3 x 3 x 3 cube, made up of 27 1 x 1 x 1 cubes, is dipped in a bucket of paint. After the paint has dried, the 3 x 3 x 3 cube is taken apart into its 27 individual cubes. How many of these individual cubes have paint on three sides, two sides, one side, zero sides? What if it were a 4 x 4 x 4 cube? What if it were a 5 x 5 x 5 cube? What if it were a 10 x 10 x 10 cube? What if it were an n x n x n cube?

Grades 10–12: 3D Tic-Tac-Toe

In a standard game of tic-tac-toe, a win occurs when 3 X's or 3 O's are all in a row (colinear). There are 8 ways to win in a standard game of tic-tac-toe—three up and down, three side to side, and two diagonally. How many ways are there to win in 3D tic-tac-toe, where the rules are the same—a win is 3 colinear X's or O's?

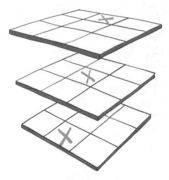

What if it were played on a 4 x 4 x 4 board, where 4 in a row are needed for a win? What if it were played on a 5 x 5 x 5 board, where 5 in a row are needed for a win? What if it were played on a n x n x n board, where n in a row are needed for a win?

CHAPTER 11

HOW STUDENTS TAKE NOTES IN A THINKING CLASSROOM

· ·

Having students take notes is one of the most enduring institutional norms that permeate mathematics classrooms all over the world. In my visits to the original 40 classrooms, I witnessed students taking notes in all of the 23 high school classrooms and in more than half of the 17 elementary classrooms. In this chapter you will first see the results of research that looks at the normative practice of note taking through the lens of thinking. You will then learn about an alternative method for having students take notes that makes it a thinking activity that is relevant for students in Grades 3 to 12.

> Having students take notes is one of the most enduring institutional norms that permeate mathematics classrooms all over the world.

THE ISSUE

When I observed students taking notes in those original 40 classrooms, they were most often in the form of what I call *I-write-you-write* notes. This is where the classroom teacher writes notes on the board, and the students write down, word for word, symbol for symbol, what the teacher has written. These notes are usually a combination of definitions and worked examples and are accompanied by the teacher's verbal explanations as they write. Occasionally, classroom notes take the form of *fill-in-the-blank* notes, where the students are provided with mostly completed notes and are asked to fill missing pieces as the teacher writes on the board.

Given that, in many instances, more than half the lesson was dedicated to this activity, it must be an important and worthwhile endeavor. When I interviewed teachers about this, the two most common reasons teachers gave for having students write notes were that (1) it created a record for them to look back at in the future, and (2) it was a way for them to learn. Although using notes as a record was suggested almost exclusively by high school teachers, using notes as a form of learning was supported by teachers from Grades 3 to 12. If writing notes is a way to enhance learning, and if thinking is a necessary precursor to learning, then students taking notes must be a thinking activity.

THE PROBLEM

To test whether students taking notes is indeed a thinking activity, we documented the various studenting behaviors exhibited during I-write-you-write note taking in 10 different classrooms (Grades 6 to 12)[1] as well as during fill-in-the-blank note taking in three classes (Grades 9, 10, 11). We also distributed a one-question survey to all the students in these 13 classes and conducted short interviews with 150 select students.

In the 10 I-write-you-write classrooms, 14% of students did not take notes at all, most commonly—they said—because it was difficult to take notes and listen at the same time. And they would rather listen. Some said they never looked at the notes anyway, and others cited things like "I forgot my notebook" or "I forgot my pencil", which we took to be proxies for "I don't want to take notes."

Of those students who did take notes, we noticed that more than half were not keeping up with the teacher. Why was this significant? To answer this, I need to differentiate between what we call *live notes* and *dead notes*. Live notes are the product of the real-time generation of notes by the teacher, where the teacher is working through an example, demonstrating the sequence of how something is to be done, and providing a narrative as the work unfolds. Whereas live notes are a chronologically linear process, they are often spatially non-linear. That is, the notes are not appearing on the board in a strictly top-to-bottom, left-to-right fashion. Much of mathematics is, by its nature, non-linear.

> Of those students who did take notes, more than half were not keeping up with the teacher.

For example, consider a teacher demonstrating how to make a graph of a function. Chronologically, the teacher would first write the function or relationship. Then they would make a table of values, perhaps generating a list of values for the x variable. This would be followed by calculating and recording the y-value for each x-value. The teacher would then draw an appropriate set of x and y coordinates with consideration of the domain and range of the values in their table of values. This would be followed by a labeling and numbering of the axes, plotting of each ordered pair from the table of values, and finally the drawing of the curve. All the while, the teacher would be narrating what is happening and why certain choices and decisions are being made and how and why certain actions are being performed.

Although linear in a chronological sense, how this is unfolding spatially is non-linear. The teacher starts in the top left hand corner of the board

---

[1] See Liljedahl & Allan (2013a) for a more detailed description of the methodology used.

or page with the function, then works vertically downward to create the table of values and list the *x*-values. This is followed by a return to the top of the table of values to fill in the *y*-value for each *x*-value. Some of the calculations for this are being performed in the bottom right of the board. The teacher then moves to the top-middle of the board to start drawing the *y*-axis, and so on. When this process is complete, the resultant static images are the *dead notes* (see Figure 11.1) in which neither the chronological nor spatial sequencing is evident.

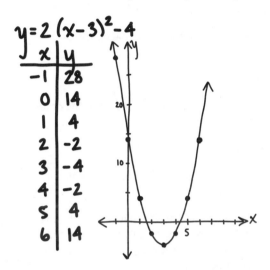

Figure 11.1   Teacher-generated notes.

Setting aside for the moment the fact that this scenario represents a situation in which the teacher, rather than the students, is the one doing all the thinking, the other problem is that the more than half of the students who are not keeping up with the teacher are left to copy dead notes. This is not like copying a sentence written on the board where they can read it, make sense of it, and write it out. Copying an image without the benefit of chronological and spatial sequencing requires students

> This continual effort to track, write, and keep up with the teachers' thinking requires a huge amount of cognitive effort, which causes students to fall further behind—to the point where they just stop listening and trying to make sense of what they are writing.

to look repeatedly between the board and their page to get all of the details of, for example, the table of values, the axes, the plotted ordered pairs, and the graph into the correct place. This continual effort to track, write, and keep up with the teacher's thinking requires a huge amount of cognitive effort, which causes students to fall further behind—to the point where they just stop listening and trying to make sense of what they are writing.

This is exactly what we observed in the behavior of the students who were not keeping up with the leading edge of the live notes. It started with a lot of head movement as the students looked from dead notes, to their notebook, to the live notes and the teacher, back to the dead notes, back to their notebook, and so on. Eventually, and sometimes rather quickly, they stopped looking at the live notes, and when this happened, they stopped trying to gain the benefit of the chronological and spatial sequencing of what the teacher was doing and saying and resigned themselves to just copying the dead notes.

Researcher So, I noticed you were falling behind on the notes today.

Philip Yeah. That usually happens.

Researcher So, how do you know what order to write the notes in if you are not keeping up?

Philip Huh? I just write what's on the board.

Researcher Of course. But what part of the board do you start with?

Philip I start with the top left and then just kind of copy what is on the board. I don't really follow any order. Sometimes I start with the biggest thing and then add the little things around it.

And when this happened, they stopped listening to the teacher.

Researcher So, when you are behind on the notes like that, are you trying to listen to what the teacher is talking about?

Stephanie Sometimes . . . I guess.

Researcher It didn't look like that at the end there.

Stephanie No. Not at the end. I had given up by then. But at the beginning when I was trying to keep up, I was listening.

This may be why several students declared that they would rather listen than write notes. For these students, writing notes and listening to the teacher are an either/or situation—either they are listening to the teacher at the leading edge of the live notes, or they are copying dead notes, like Stephanie did, without listening.

Interviewer Say a bit more about why you can either listen or you can write notes. Why not both at the same time?

Alana	Oh I can do both for a while—at the beginning. But when I start to fall behind I just end up writing the whole class and miss everything the teacher is saying. And, in the end, I don't even know what I have written anyways.
Interviewer	Because you weren't listening?
Alana	No. Because I actually stop paying attention to what I am writing. I just kind of go into this zombie state and just copy what is on the board. This is why I have decided to just listen to the teacher. I can take a picture of my friends' notes later.

Like Alana, nearly all of the students we observed either falling behind on notes or only listening had the same refrain—*copying dead notes was a mindless activity*. This was evident in the body language of the students we observed. As they fell behind they slipped into a sort of malaise as their attention to what they were doing waned.

> Copying dead notes was a mindless activity.

Fill-in-the-blank notes seems like a good way to mitigate the issue of students falling behind. Indeed, when asked why teachers use fill-in-the-blank notes, all three teachers said that this is a way to make sure that students can spend more time listening than writing. However, when we observed student behaviors in these classes, something new and even more troubling emerged. Although it is true that the students spent much less time writing, and that they did not fall behind, very few students (35%) actually spent the time listening—at least in the ways the teachers had intended. Instead, they were listening for key terms or phrases, or watching for key steps in the examples. In fact, many of the students were observed to be not listening at all and only copying what their neighbors wrote into the blanks. They were simply trying to get the information necessary to fill in the blanks rather than trying to understand what the bigger picture was.

This is not to say that all of the students who were keeping up were any more actively involved than those who weren't. Of all of the students we interviewed, just over a third *were* keeping up with the live notes, but most of them still admitted that they weren't thinking about much.

Interviewer	Not much! Really?
Samantha	Yup.
Interviewer	So, what are you doing the whole time you are taking notes?

Samantha	Just copying them down. I actually like this. It's easy and I don't need to think much.

> Both I-write-you-write and fill-in-the-blank methods of producing notes were activities that were antithetical to student thinking—and antithetical to building a thinking classroom.

The only difference between these students and those copying dead notes was that they were just copying faster. All of this showed us that the time spent having students create notes was, for the most part, not well spent. Both I-write-you-write and fill-in-the-blank methods of producing notes were activities that were antithetical to student thinking—and antithetical to building a thinking classroom. The next question, then, became—what was the value of these notes given the non-thinking time it took to create them? What did students use them for?

To answer this question, we asked each of the students in these 13 classes to answer a one question survey—*Do you use your notes on a regular basis? Please comment.* The results of this survey were disappointing, with only 18% of students (five or fewer students in most classes) stating that they regularly used their notes in one way or another.

Researcher	So, let me get this straight. I just saw you write notes for 35 minutes, and you are telling me that you won't use them? Why not?
Nahal	I mean, they're almost exactly what's in the textbook. So, if I need to look something up while doing my homework, I'll just flip back a few pages in the textbook.
Researcher	So, why write notes?
Nahal	It's what we do in class.
Researcher	Can you explain a little bit more about why you don't use your notes when doing homework?
Steven	I try to do the homework in my spare, which is the next block. So, I can usually still remember things from class.
Researcher	So, why write notes?
Steven	I don't know. I guess it's just what we do.

So, if the writing of notes is a non-thinking activity, and few students even used them, the question then became—how can we reimagine note taking as a meaningful thinking activity?

TOWARD A THINKING CLASSROOM

To figure out how to make note taking a thinking activity, I began by observing students in settings where they were not required to take notes but did anyway. Here I asked people who were not writing notes, why they were not—especially given that some of their peers were. I learned there are three main reasons that students don't take notes if not required to.

1. They will not write notes on something that they do not find interesting or important.

2. They will not write notes on something they know they can find elsewhere—like a PowerPoint presentation, an article, or the textbook.

3. They will not write notes on things they think they will remember.

What this means, is that students in these situations only take notes about things they feel are important and for which documentation does not exist somewhere else, and for their future forgetful selves. This is the essence of mindful (as opposed to mindless) and meaningful (as opposed to meaningless) note taking and became the focus of our first intervention into note taking—we were going to have students write *notes to their future forgetful selves*. In the end, we discovered that this was a *thinking approach* to note-taking—in other words we found a way to make meaningful notes. However, this didn't come easily to all students at first.

> *Notes to their future forgetful selves* is the essence of mindful (as opposed to mindless) and meaningful (as opposed to meaningless) note taking.

We tried this thinking approach to note taking with 11 teachers (Grades 4 to 12) in 48 classrooms. These teachers were all working within the framework of having students do curriculum-based thinking tasks in random groups on VNPSs, and their rooms were all *defronted*. They were all at varying stages of competency with the rest of the aforementioned thinking classroom practices. The teachers each committed to running their classrooms as they always did, with the exception that towards the end of the lesson they would ask their students to sit down in their desks to write some notes to their future forgetful selves. They elaborated on this by asking, "What are you going to write down now so that, in three weeks, you will remember what you learned today?" I asked that they allow a minimum of 10 minutes for this activity.

This worked wonderfully in the four Grade 4, 5, and 6 classes, with students producing beautiful, and personalized, representations of what they had learned. Some students used a lot of pictures, some included examples, and some wrote out sentences explaining what they had done. Of course, some students struggled, but they were in the minority. It worked OK in the Grade 11 and 12 classes as well, but these students struggled with this more than the younger students. Suggestions that they include examples, and annotate those examples with comments about what they did, helped. It also helped to share examples of different types of notes and to ask students to discuss, in random groups, which notes were useful and why. This helped make explicit the qualities of notes that were meaningful to them. What helped the most, however, was circling back three weeks later with questions that required them to use these notes. This gave relatively quick feedback to them as to what they would find useful three weeks down the road. By and large, they had underestimated how much they would forget, and this experience had an immediate and significant impact on how carefully they annotated new notes from that point forward.

Interestingly, the Grade 8 and 9 students struggled the most with this task. The vast majority of these students were at a complete loss as to what to write and for what purpose. Some wrote everything they could think of, while others were paralyzed by the vast possibility of what to write. Suggestions to include examples helped for some, but only exasperated those who were struggling with what to select. The use of exemplars and circling back three weeks later had little impact on these students. Interviews with these students revealed some of the issues.

Researcher   I see that you haven't written much.

Patrick   Hmm . . .

Researcher   Why not? What's the problem?

Patrick   I don't know. In my other classes we just copy what the teacher writes on the board. I like that better.

Researcher	Why is that?
Patrick	I don't know. I don't have to think so much about it, I guess. Like, I don't have to decide what to write. That's hard.

What we were finding, in general, was that by Grade 8, students have become encultured into a more mindless form of note taking, and they have lost the ability to decide for themselves what to write down. The younger students did not suffer from this and were more than happy to write what they wanted. This is not to say that they did not benefit from some guidance, but they were not impeded by a preconceived notion that taking notes is a mindless activity.

> By Grade 8, students have become encultured into a more mindless form of note taking and they have lost the ability to decide for themselves what to write down.

Likewise, the older students, although even more encultured into note taking as a mindless activity, were more aware of the role of notes as a potentially meaningful activity. The Grade 8s and 9s had not yet come to this realization as they considered note taking something they did for the teacher, not for themselves.

What the Grade 8 and 9 students needed, in essence, was a support tool to help scaffold and organize their notes. With this in mind, we turned to graphic organizers. Graphic organizers have been used with great success in humanities teaching for decades. The initial impetus to try using these as a template for note taking came from a group of teachers who had some experiences teaching in this area.

> What the Grade 8 and 9 students needed, in essence, was a support tool to help scaffold and organize their notes. They needed graphic organizers.

Graphic organizers can take many forms, but we experimented with four different types—the first of which (Type I) is just to have students write their meaningful notes in a cell of limited size (see Figure 11.2). The idea is that they can write whatever they want in their notes today, but it all has to fit into a cell that is, in the case of the example, one-eighth of a page. This constraint on available space helps students focus what they want to write to their future forgetful selves. As before, some students chose to use diagrams, or examples, while others opted more for words.

Figure 11.2 Type I: Graphic organizer with cells as restrictions.
Source: Courtesy of students at St. Mother Theresa School.

The second type of graphic organizer (Type II) has the same sort of cell structure as Type I, but students are allowed to use those cells as a way to organize different aspects of their notes (see Figure 11.3). So, whereas the limited space organizer (Type I) dictates where and how much the students can write, in Type II neither of these is limited. Rather, the cells are a way for students to demarcate different aspects of their notes.

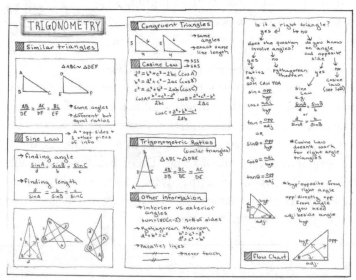

Figure 11.3 Type II: Graphic organizer with cells as demarcations.

LR1 Data Management		Linear Relations	
Scatter plot		Line of best fit	
Collect, organize, & analyze data		Describe trends & relationships in data	

LR2 Characteristics of Linear Relations			Linear Relations
Linear	Non-linear	First differences	
Line/Curve of best fit	Rate of change	Initial Value	
Direct variation	Partial variation	Create tables of values, graphs, & equations	

Figure 11.4 Type III: Graphic organizer with prelabeled cells to demarcate different subtopics.

Figure 11.5 Student fills in a Type III graphic organizer with prelabeled cells to demarcate different subtopics.

Type III (see Figures 11.4 and 11.5) formalizes the organization and demarcation aspects of the Type II graphic organizer. Unlike the Type II notes, however, with this graphic organizer the teacher prelabels the cells according to the things that they feel are important for students to record. The student can write what they want in these cells, but the category of what goes in those cells is predetermined.

What all three of these organizers have in common is a focus on brevity and, although quite different in style, result in an entire unit of notes on two sides of a page. This, along with the allocation of where to write, seems to be tremendously liberating for students who are paralyzed by the possibility of choice.

> Through worked examples, students have the potential to communicate to their future selves not only *how* to do something, but *why* something is done.

Something else they have in common, but not explicitly, is that they all allow for the possibility of students including worked examples. Unfortunately, our research showed that very few students seize this opportunity on their own. This is problematic in that worked examples are an important part of notes and convey what John Mason and David Pimm (1984) refer to as *the general in the particular*. Through worked examples, students have the potential to communicate to their future selves not only *how* to do something, but *why* something is done. This communication is enhanced, of course, if they go beyond the inclusion of worked examples and begin to annotate these examples. To make this explicit, the Type III (see Figure 11.4) graphic organizer could include demarcated spaces for worked examples to be added.

Alternatively, an altogether different type of graphic organizer can be used (Type IV). As opposed to breaking content into subtopics (types II and III), this organizer differentiates among vocabulary or definitions, big ideas or concepts, procedures, and examples (see Figure 11.6). These cells can be created and prelabeled by the teacher or by the students. Some teachers have students create and fill a new one of these for every lesson, while others have the students do one per unit and add to it at the end of each lesson. The headings are not fixed and can be varied according to what an individual teacher feels is important for students to record. What is important, however, is that the inclusion of worked examples is explicit.

Vocabulary/Definition	Big Ideas/Concepts
Procedures	Examples

Figure 11.6 Type IV: Graphic organizer with prelabeled cells to demarcate different aspects of a topic.

Regardless of whether graphic organizers are used or not, we learned from experimenting with having students write notes to their future forgetful selves that there are three distinct competencies needed for students to be able to produce useful worked examples that are meaningful to them:

- Creation: producing a worked example.

- Annotation: using small phrases and side examples as signposts on the journey through a worked example.

- Selection: choosing an appropriate question to form the basis of a worked example.

> There are three distinct competencies needed for students to be able to produce useful and meaningful (to them) worked examples: *creation, annotation,* and *selection.*

Of these, creation is the easiest competency to acquire. Students, when given a specific question, usually had little difficulty producing the worked example. Very few students, however, spontaneously used this as an opportunity to begin to annotate their examples. We found that we could enhance this skill in students by first having them annotate complete, but incorrect, worked examples.

The most difficult of the three competencies, however, is the selection of a question to create a worked example in response to. As teachers, we know that a worked example needs to stand, in a general sense, as a proxy for a wide range of examples. Whereas an appropriately chosen question can make transparent every step of the worked example, poorly chosen numbers can obfuscate what is happening. For example, $2^2 = 4$ makes ambiguous the role that the exponent is playing in this equation—is it a multiplicand or the number of times 2 is multiplied?

Students' selection competency improved when teachers gave them a set of questions to choose from. This occurred in two ways. The first was to give the students a list of questions from which to choose one or two to develop as worked examples. This could either be integrated right into a Type III or Type IV graphic organizer or simply listed on the board. The second was to draw their attention to the fact that they had just worked through several curriculum questions in their random groups on the whiteboards and that one or two of these may serve as potentially good worked examples. Regardless, narrowing the choice to some small finite number helped students to begin to make appropriate selections.

In the end, we studied teachers' efforts to scaffold students' autonomous note taking in five classrooms (Grades 5–11), using some variation of the aforementioned graphic organizers alongside strategies for helping students to create or select and then annotate worked examples. We now observed 75%–100% of students taking notes, depending on the class, and 50% of students referring back to their notes at some point. Often this referring back occurred when they started doing their check-your-understanding questions (Chapter 7), but it also occurred during their time working on whiteboards in random groups when, as a group, they felt the need to look up something they had previously learned. The students who took notes in this mindful and thoughtful way found more meaning in, and use for, them.

> The students who took notes in this mindful and thoughtful way found more meaning in, and use for, them.

These notes became a way for students to consolidate the learning that they had experienced within their collaborative groups. Unlike the teacher-led consolidation discussed in Chapter 10, meaningful note taking is student led and is a vital part of transitioning from collective knowing and doing into individual knowing and doing. This will be discussed more in Chapter 15.

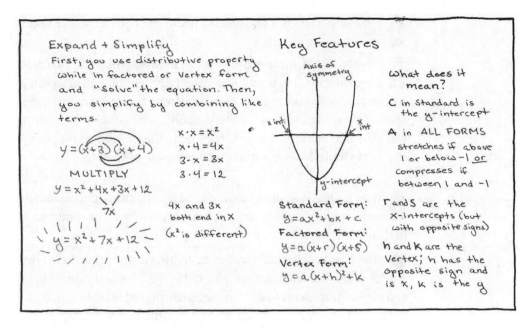

Figure 11.7 Example student notes.

Q You mention that in the I-write-you-write and fill-in-the-blank notes, students take notes for the wrong reason and then don't look at them. How can I make sure that this doesn't happen when I switch to meaningful notes?

A The most important thing you can do for this is messaging. The message that needs be used repeatedly has to do with who notes are for—they are *by them* and *for them*. This cannot be overstated. When you first introduce the notion of notes to their future forgetful selves, use this phrasing, and use it every time you discuss notes with the students. This is needed to counteract the current student belief that notes are *by the teacher—for the teacher*. So, for example, if a student asks you if their notes are good, resist the urge to impose your own judgement on the notes. Instead, ask the student if they think the notes will be useful for them in a few weeks' time.

> The message that needs be used repeatedly has to do with who notes are for—they are *by them* and *for them*.

Q How important is it for students in different grades to take notes?

A As mentioned, meaningful notes serve two purposes: (1) to create a record and (2) to reify and consolidate their individual learning. Although student notes as a record may only feel relevant for higher grades, meaningful notes as individually consolidation of learning was shown in our research to be relevant in the learning of students as young as Grade 3. This will be discussed more in Chapter 15.

Q It seems like it would be easiest to have students write meaningful notes right after I have consolidated from the bottom. Is that correct?

A Absolutely—especially when you are first introducing these types of notes. Students' meaningful notes are a natural extension of, and easier for students to do if immediately preceded by, the collective reifying experience of consolidation from the bottom (Chapter 10). For this reason, when consolidating from the bottom, mark up boards that you are discussing with circles and annotations. Also number the examples you discuss as you are doing it. These two acts will help draw students' attention to things that they may wish to include in their meaningful notes.

Q My students take pictures of work they produce in their random groups while at their VNPS. Is this the same as notes?

A We had the same question. It turns out that the answer is no. Having the pictures in their phone is not, in and of itself, very useful. A picture is just a record of what they have done. It is the act of reifying these pictures into meaningful notes that helps move the thinking and learning forward. So, if students are wanting to take pictures of their work, or the work of others, encourage them to do so. But then make sure you, likewise, encourage them to turn those pictures into notes to their future forgetful selves.

Q Do you have any other hints to help my students produce better meaningful notes?

A A very easy, and effective, way to introduce meaningful notes, after consolidation from the bottom, is to send the students, in their random groups, back to their boards to collaboratively write notes to their future forgetful selves. Not only does the ensuing conversation drive reification, but using the collaborative large space is a very safe way to start taking notes. It also affords the immediate ability for you to help draw students' attention to notes that may be useful to them. For example, when the activity is done, give every student three sticky notes, and ask them to walk around the room and place the sticky notes on the collaborative notes that they would find most useful to them in three weeks' time. Follow this up with a discussion of what it is about the boards that received lots of sticky notes that makes them such meaningful notes, create a list, and post this list, maybe along with some pictures, the next time you do the activity. In very short order students will be writing meaningful notes on their own.

 If these notes are by students—for students, then it sounds like I can't be checking to make sure they have written something that is complete and useful to them in the future. Is that correct?

 Correct. Despite the effectiveness of meaningful notes, there will still be 15% to 30% of students who write nothing. As with check-your-understanding questions (Chapter 7), any efforts to reduce this number had an immediate negative effect on the way the rest of the students in the class engaged in the process. These notes, by their very nature are taken by students—for students. We are providing them with opportunities and structure to do what is best for them. If they choose not to seize these opportunities or use these structures, that is their choice. When we tried to force these upon them by checking or marking that they were doing them, the dynamic immediately shifted back toward what we were observing with the fill-in-the-blank notes—students mindlessly filling in boxes, copying off each other, and not using the notes—and not thinking. This shift in behavior was not just among the students we were trying to impact. This behavior spread immediately to most of the students who had previously been, with or without the graphic organizers, mindfully and meaningfully taking notes to their future forgetful selves. The notes were no longer by them—for them, they were now for the teacher. Like check-your-understanding questions, meaningful notes are an incredibly sensitive structure. Whether you choose to use graphic organizers or not, the messaging around the fact that these are by students—for students is vital. Anything you do to force the issue changes, in profoundly negative ways, who they perceive the notes are for.

 If the students choose what to write, and I can't mark their notes, how can I know that what they are writing is correct?

 You can't know that. But, then again, you can't really know whether students are doing it correctly when writing the I-write-you-write and fill-in-the-blank notes either. You just have to trust that they will.

Q How can I know that the notes to their future forgetful selves my students are writing will be useful to them in the future?

A Three weeks after you have started doing notes to their future forgetful selves with your students, give them tasks to do, either in their random groups or on their own, that will require them to draw on the knowledge that they have written down in their notes. This feedback loop is vital in helping them to understand what level of detail they need to include in their notes as a record.

Q Are graphic organizers just for Grade 8 and 9 students?

A No. We explored these initially in the context of Grades 8 and 9 because they were too encultured to note taking as being a mindless and meaningless activity, and the prompt to write notes to their future forgetful selves was not enough to override this enculturation. But, we quickly learned that graphic organizers of any type were useful for students from Grade 3 to Grade 12.

Q Should we be pulling back on the scaffolding in the graphic organizers as the students move through the year?

A Depending on the grade, yes. For example, your Grade 12 students, in preparation for postsecondary studies, should be able to write meaningful notes without any scaffolding by the end of the course. For the rest of the grades you can proceed as you feel is best. The bottom line is that notes to their future forgetful selves do not need as much scaffolding once the students realize that notes are *by them—for them*. The same is true of the scaffolding around worked examples. Although you may start the year by giving them a list from which to choose a question to develop into a worked example, the goal is to gradually move toward them selecting, for themselves, what would make a good example.

Q Can't we just stop doing notes altogether and just put them online?

A We tried this. On the plus side, this freed up more time in the lesson for thinking activities such as collaboration, discussion, and solving thinking tasks. On the minus side, however, the elimination of notes altogether also eliminated the opportunity for students to reify and consolidate their individual learning.

SUMMARY

MACRO·MOVE

☐ Have students write meaningful notes.

MICRO·MOVES

☐ Emphasize that meaningful notes are BY THEM–FOR THEM.

☐ Prompt students to write NOTES TO THEIR FUTURE FORGETFUL SELVES.

☐ Use graphic organizers.

☐ Have students collaboratively write NOTES TO THEIR FUTURE FORGETFUL SELVES.

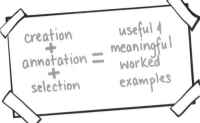

creation + annotation + selection = useful & meaningful worked examples

☐ Emphasize the importance of worked examples:
 ☐ Give students choices of worked examples.
 ☐ Have them correct incorrect worked examples.
 ☐ Make sure to annotate.

☐ Give tasks 3 weeks later that require students to use their meaningful notes.

QUESTIONS TO THINK ABOUT

1. What are some of the things in this chapter that immediately feel correct?

2. Which of your students do you think will have an easy time doing meaningful notes? Which will need more support? Which will not do them?

3. Which graphic organizer, if any, do you think would be best for your students?

4. How do you feel about the fact that if you try to manage meaningful notes, students will start to do them for the wrong reason?

5. Think of a time where you, yourself, took I-write-you-write and/or fill-in-the-blank notes. How engaged were you?

6. What are some of the challenges you anticipate you will experience in implementing the strategies suggested in this chapter? What are some of the ways to overcome these?

 TRY THIS

The following are tasks that require multistep solutions and solution pathways. As such, they are ideal for students to write meaningful notes on.

Grades K–5: Dot Patterns

Consider the following dot pattern. There are 25 dots. Find multiple ways to show that there are 25 dots by circling pieces of it and writing the appropriate number sentence.

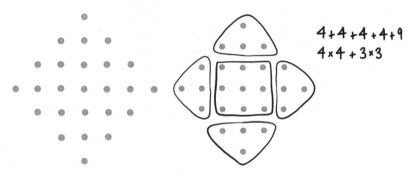

Grades 6–9: 1,001 Pennies

There are 1,001 pennies lined up on a table. Starting at one end of the line, I replace every second coin with a nickel. I then go back to the beginning and replace every third coin with a dime. Finally, I go back to the beginning and replace every fourth coin with a quarter. How much money is now on the table?

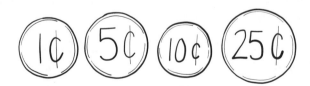

Grades 10–12: Bank Robber

A bank robber is being chased by a bank guard when they fall into a square swimming pool. By the time the guard gets to the corner of the pool, the robber has swum into the exact middle of the pool. The guard can run faster than the robber can swim. But the robber can run faster than the guard can run. What direction should the bank robber swim in (in a straight line) to maximize their chance of escape?

CHAPTER 12

WHAT WE CHOOSE TO EVALUATE IN A THINKING CLASSROOM

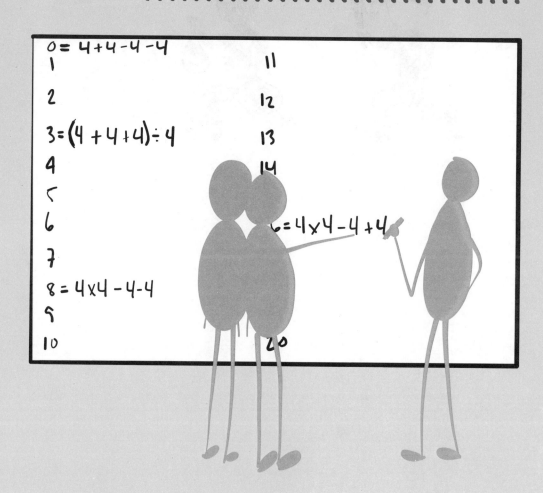

In Chapter 8 we looked at the role that fostering autonomy can play to help mobilize knowledge, which, in turn, helps a thinking classroom to function better. In this chapter you will learn about how we, as teachers, can use evaluation to further develop and refine these, and other, student competencies. By the end of this chapter you will be able to cocreate—with students—and administer rubrics for improving student perseverance, risk taking, autonomous actions, et cetera.

THE ISSUE

What competencies are valuable for students to be successful in a thinking classroom? Take a moment to answer this question. Now pick the three competencies on your list that you think are the *most* valuable for student success in a thinking classroom.

Over the last few years I have asked teachers in over 50 different professional development settings this exact same question. Working in random groups of three and on vertical non-permanent surfaces, the lists that they produce usually contain between 10 and 20 different competencies ranging from curiosity to critical thinking to patience. When I ask each group to choose the three competencies that they think are most valuable—those most likely to lead to student success in a thinking classroom—and then compile every group's top three competencies into a master list (a union of the top threes), an interesting thing happens. Regardless of geography, grade levels, or professional development setting, the same three competencies appear every time:

- perseverance
- willingness to take risks
- ability to collaborate

Even when I ask the question without situating it in a thinking classroom setting—*what competencies are valuable for students to be successful at the next task/the next lesson/the next unit/mathematics—*the same three competencies emerge every time. Hundreds of teachers—maybe including you—have identified these as the competencies most valuable to student success in not only thinking classrooms, but mathematics classrooms in general. This is not to say that these are the only three competencies that end up on the master list. Curiosity, autonomy, self-responsibility, grit, positive views, self-efficacy, et cetera will also make appearances, but not with the same consistency as perseverance, willingness to take risks, and ability to collaborate.

My next question to these teachers, and to you, is always the same— *Is it our job, as teachers, to wait for students to come to us with these competencies in place, or is it our job to develop these competencies within the students that we have in front of us?* The answer is always the same—*it is our job to develop these competencies.* Of course, it is. In fact, many jurisdictions around the world have these, or similar, competencies (sometimes called processes), as explicit outcomes of their curricula.

THE PROBLEM

Whether or not we should develop these competencies is not the question we should concern ourselves with—the answer is clear. You are possibly already doing this through problem-based learning, collaborative groups, or mathematical discussion. If you have been implementing the thinking practices as you read along, you are definitely beginning to develop these competencies.

> We need to put our evaluation where our mouth is. We need to start evaluating what we value.

The question we should concern ourselves with is— *How are you going to evaluate them?* Depending on the mandates of the curriculum you work under, you may neither be required to, nor have a desire to, evaluate such competencies. But this misses the point. If these competencies are so valuable, then we *need* to evaluate them—and *how* we evaluate them becomes the key question. This is because evaluation is a double edged sword. When we evaluate our students, they evaluate us—for what we choose to evaluate tells our students what we value. So, if we value perseverance, we need to find a way to evaluate it. If we value collaboration, we need to find a way to evaluate it. No amount of talking about how important and valuable these competencies are is going to convince students about our conviction around them if we choose only to evaluate their abilities to individually answer closed skill math questions. We need to put our evaluation where our mouth is. We need to start evaluating what we value.

This is not to say that we stop evaluating students' abilities to demonstrate individual attainment of curriculum outcomes, but we need to also find ways to evaluate those things that we actually say we find most valuable. If you have been implementing the 11 practices presented thus far, you are starting to feel this tension—the feeling that somehow we need to broaden our assessment practice to begin to value the competencies that we are espousing, developing, and utilizing every day in a thinking classroom. But this need, in and of itself, only tells us what we want to do. It doesn't tell us how to do it—and *how* we do it turns out to be very important.

TOWARD A THINKING CLASSROOM

This tension was the exact starting point of the research into how we can begin to evaluate what we value—to begin to find ways to value the perseverance, collaboration, and willingness to take risks that we were asking students to undertake. Clearly, your typical test is not going to provide the means to achieve this. We needed other tools, other metrics, other instruments that could evaluate complex competencies. One thought was to try to leverage a tool that already existed—*a rubric*—to see if it could help us.

> We needed other tools, other metrics, other instruments, that could evaluate complex competencies.

The Rubric Approach

You may be familiar with or have used rubrics—or matrices as they are sometimes called—like the one in Figure 12.1. This rubric, although not designed to evaluate the competencies we were interested in, was widely used within the jurisdiction where the research was taking place and was familiar to both teachers and students. As such, it became a starting point into our research into the use of rubrics. And like our research into homework (Chapter 7) and student notes (Chapter 11), our research into rubrics began with a deep dive into how effective the status quo was. And the status quo was not good.

Figure 12.1 An existing four-column rubric (British Columbia Ministry of Education, 2020).

Aspect	Not Yet Within Expectations	Meets Expectations (Minimal Level)	Fully Meets Expectations	Exceeds Expectations
Snapshot	*The student is unable to meet basic requirements of the task without close, ongoing assistance. Unable to provide a relevant extension.*	*The work satisfies most basic requirements of the task, but it is flawed or incomplete in some way. May produce a simple extension with help.*	*The work satisfies basic requirements. If asked, the student can produce a relevant extension or further illustration.*	*The work is complete, accurate, and efficient. The student may volunteer an extension, an application, or a further illustration.*
Concepts and Applications* • recognizing mathematics • grade-specific concepts, skills • patterns, relationships	• unable to identify mathematical concepts or procedures needed • does not apply relevant mathematical concepts and skills appropriately; major errors or omissions • often unable to describe patterns or relationships	• identifies most mathematical concepts and procedures needed • applies most relevant mathematical concepts and skills appropriately; some errors or omissions • may need help to describe and use patterns and relationships	• identifies mathematical concepts and procedures needed • applies mathematical concepts and skills appropriately; may be inefficient, make minor errors or omissions • describes and uses basic patterns and relationships	• identifies mathematical concepts and procedures needed; may offer alternatives • applies mathematical concepts and skills accurately and efficiently; thorough • independently describes and uses patterns and relationships
Strategies and Approaches • procedures • estimates to verify solutions	• appears unsystematic and inefficient • results or solutions are often improbable	• generally follows instructions without adjusting or checking • may need reminding to verify results or solutions; estimates are generally logical	• follows logical steps; may be inefficient • makes logical, relatively accurate estimates to verify results or solutions	• structures the task efficiently; may find a shortcut • makes logical estimates to verify results or solutions
Accuracy • recording, calculations	• often includes major errors in recording or calculations	• may include some errors in recording or calculations; generally "close"	• recording and calculations are generally accurate; may include minor errors	• recording and calculations are accurate; may use mental math
Representation and Communication • presenting work • constructing charts, diagrams, displays • explaining procedures, results	• work is often confusing, with key information omitted • often omits required charts, diagrams, or graphs, or makes major errors • explanations are incomplete or illogical	• most work is clear; may omit some needed information • creates required charts, diagrams, or graphs; some features may be inaccurate or incomplete • explanations may be incomplete or imprecise	• work is generally clear and easy to follow • uses required charts, diagrams, or graphs appropriately; may have minor errors or flaws • explains procedures and results logically in own words	• work is clear, detailed, and logically organized • uses required charts, diagrams, or graphs effectively and Accurately • explains procedures and results clearly and logically; may include visuals

Cursory observations in the classrooms when rubrics like the one in Figure 12.1 were in use revealed that students didn't look at the feedback the rubrics were providing. In one lesson I was observing, 75% of students spent less than 10 seconds looking at the rubric when it was returned to them. The rest of the students spent less than one minute looking at the feedback. Irrespective of how much time was spent grading student work, and irrespective of how carefully teachers highlighted comments on the rubric to match what they were seeing in the student work, most students were not looking at, never mind attending to, the feedback. The messages were being delivered—they just weren't being received. Therefore, these rubrics were having little impact on student behavior. And we wanted rubrics that had impact.

> In one class, 75% of students spent less than 10 seconds looking at the rubric when it was returned to them.

So, we began a process of successive designing, modifying, and testing various rubrics. Our goal was simple—find a rubric that could be used to increase thinking behaviors such as perseverance, collaboration, and willingness to take risks. What emerged from this research was a rubric that looked very different from the one in Figure 12.1 (see Figure 12.2).

> What emerged from this research was a rubric that looked very different.

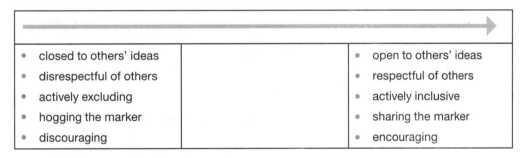

• closed to others' ideas		• open to others' ideas
• disrespectful of others		• respectful of others
• actively excluding		• actively inclusive
• hogging the marker		• sharing the marker
• discouraging		• encouraging

Figure 12.2 Collaboration rubric.

There are five visible differences between this new rubric and the ones we started with.

> It is an *observational rubric* to be used by the teacher *while students are thinking*—not after they are finished thinking.

1. FOCUS

First, whereas the initial rubrics looked for competencies in student *work*, this new rubric looks for competencies in student *actions*. That is, it is an *observational rubric* (Elrod & Strayer, 2015) to be used by the teacher *while students are thinking*—not after they are finished thinking. This, it turned out, gave us much more direct access to student competencies and, in return, produced bigger changes in student behaviors.

2. NUMBER OF COLUMNS

> The problem is that we cannot nuance language well enough to communicate differences among that many columns. The bigger problem is that we think we can.

Most of the existing rubrics we initially looked at consisted of at least four columns—and in some cases as many as five or six. The problem is that we *cannot* nuance language well enough to communicate differences among that many columns. The bigger problem is that *we think we can*. In truth, however, that nuancing garbles the feedback for students.

Researcher	So, what do you need to improve on for next time?
Shen	I don't know.
Researcher	Why not?
Shen	Like, sometimes I feel like I am mostly right, but my teacher thinks I might have some errors.

Shen's last comment refers to the accuracy row in Figure 12.1, but could apply to several places in that—or any four-column—rubric. In our attempts to differentiate student work across four columns, it is easy to create language for the first column. This column generally describes poor work, and that is easy to articulate. Likewise, the good work represented in the fourth column is easy to clearly explain. It is when we try to differentiate between the two columns in the middle that we get into trouble and end up trying to say that *mostly right* is different from *might have some errors*, or *primarily* rather than *usually*, or *sometimes* rather than *occasionally*. Even if we believe it is clear, it is not clear to the students. They can see that they live in different columns, but they don't know what they need to do to move from one to the next. Put simply, the information is too much and too nuanced to be useful to them. Like anyone, they were more likely to actually read and respond to something that is more direct, more digestible, and more visual.

Ironically, we discovered that information in rubrics like the one in Figure 12.1 was ambiguous to teachers as well. In many cases where we secretly scrambled a rubric, teachers were unable to correctly unscramble it—sometimes even scrambling it more. It turns out that that level of nuance was often just as difficult for them to manage and interpret as it was for students. They just didn't have a better model to follow. Once we figured this out, we moved to three-column rubrics, and all of those problems went away. The rubrics became easier to create and use, and students found them easier to interpret—they spent more time looking at the feedback, and they were able to take action on that feedback.

This was true for students in Grades 2–12. For K–1, however, we found that even three columns were initially too many. Students in these grades are still developing their ability to see and sense nuance and subtlety, and they are still experiencing their world through a lens of binary opposites—good-bad, high-low, hot-cold, wet-dry, big-little, and so on (Bettelheim, 1976, Egan 1988, Zazkis & Liljedahl, 2008). Therefore, for these grades the rubrics were only two columns and were constructed, for the most part, using visuals (see Figure 12.3).

Figure 12.3: K–1 collaboration rubric.

3. HEADINGS

The third difference is that the headings at the top of each column have been replaced by an arrow. Interviews with students revealed that many of them were seeing the headings in the initial rubric as descriptions of *who* they are, rather than *where* they are. Deeper investigation revealed that how students interpreted these headings was influenced by whether they had a growth mindset or a fixed mindset (Dweck, 2016). Students with a growth mindset saw these labels as descriptors of *where* they were, while students with fixed mindsets saw them as *who* they were—even when temporal or positional language such as *not yet* or *on the way* were being used. This is a problem. By replacing this language with the arrow, even students with a fixed mindset began to see the feedback as a descriptor of *where* they were.

> Many students were seeing the headings in the initial rubric as descriptions of *who* they are, rather than *where* they are.

4. REDUCTION IN LANGUAGE

The fourth visible difference is the absence of language in the middle column. When we were working with three-column rubrics with language in each column, the students treated each column as a discrete level of the competency being evaluated. So, when they were asked to self-evaluate their group's competency, they highlighted their performance as fitting in one of the three columns (see Figure 12.4). This is how they had been evaluated on four-column rubrics, so it made sense that they would do the same on a three-column rubric. When we removed the language in the middle column, however, their way of self-evaluating completely changed. Students no longer thought of the rubric as a collection of discrete levels, but rather as a continuum with the first and third columns acting as endpoints, and the way they highlighted reflected this (see Figure 12.5). Even students who had previously used three-column rubrics for self-evaluation almost seamlessly switched to seeing this as a continuum.

> Students no longer thought of the rubric as a collection of discrete levels, but rather as a continuum.

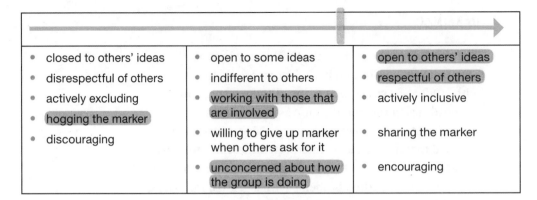

Figure 12.4 Highlighted collaboration rubric with language in all three columns.

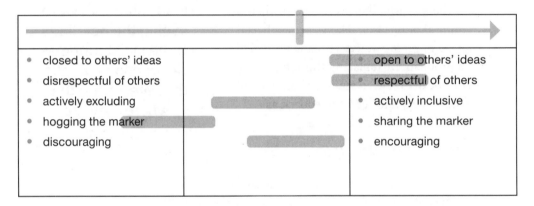

Figure 12.5 Highlighted collaboration rubric with language in two end columns.

5. **Reduction in Number of Competencies Assessed**

You may have also noted there was not only a reduction in language in general, but a reduction in the number of competencies being assessed in a single rubric. Whereas the four-column rubrics we were seeing tended to try to evaluate three, four, or even five competencies at a time, the rubrics in Figures 12.2 and 12.3 only evaluate one. And it does so using indicators that are more straightforward and approachable to students. This is because of how this rubric was created and used.

Creating the Rubric

Most of the rubrics we were initially looking at were created by teachers, teaching organizations, publishers, or ministries/departments of education. The rubrics in Figures 12.2 and 12.3 were created with and by the students who were going to be evaluated by them. This is not a new idea.

The practice of having students coconstruct success criteria, scoring guides, and rubrics has been around for a long time (Davies & Herbst, 2013; Staples, 2007). The purpose of coconstructing a rubric is the ownership that happens when students see that they have a voice in what will be evaluated and how they will be evaluated. Coconstruction of rubrics allows for the emergence of language and terminology that, although unique and potentially idiosyncratic, is clear to the students who had a hand in creating it. That doesn't mean we can't look at other instruments for inspiration and guidance, but we need to go through the process of coconstruction with our students. Our research on using rubrics in thinking classrooms confirmed this over and over again.

> Coconstruction of rubrics allows for the emergence of language and terminology that, although unique and potentially idiosyncratic, is clear to the students who had a hand in creating it.

In the thinking classroom setting, the way this looks is amazingly streamlined and time efficient, and it begins at the end of a lesson where you see a competency lacking.

Teacher	There were some really interesting answers emerging from your group work today. But I noticed that a lot of you were giving up quite easily. You didn't have very good perseverance. So, what I would like to know is, what does good perseverance look like? [Teacher draws a T-chart on the board and writes GOOD on the top of the right column (see Figure 12.6).]
Student	Not giving up when it gets tough. [Teacher writes this under the GOOD heading.]
Student	Looking around when you get stuck.
Teacher	Looking for what?
Student	A hint. Looking around for a hint when you are stuck. [Teacher writes this on the board.]
Student	Asking the teacher for help. [Teacher writes this on the board.]
	. . .
Teacher	Ok. So, what does bad perseverance look like? [Teacher writes BAD at the top of the left column.]
Student	Giving up right away.
Teacher	Right away?

Student	Ok. Just giving up.
Teacher	When?
Student	As soon as we get stuck or it gets hard. [Teacher writes this on the board.]

. . .

What is interesting about this process is that no matter how lacking a class may be in demonstrating an observable competency, they can always generate a list of indicators of what it would look like to have that competency. In the hundreds of times this has been tried, this has always been true—even if students don't exhibit the desired behavior, they know what the desired behavior looks like.

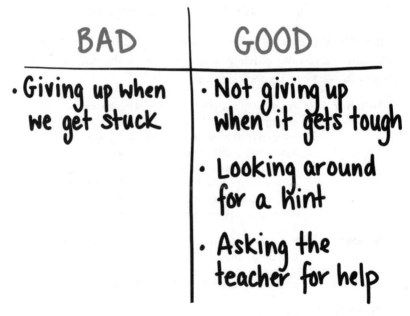

Figure 12.6 Coconstructed T-chart for developing perseverance rubric.

Before the next class, the indicators in this T-chart are then used by the teacher to build a rubric like the one in Figure 12.2 or 12.3—depending on the grade. This is not to say that we take all of their suggested indicators, or that we keep the order the same. But, what we do take, we take verbatim from what is in the T-chart. As you see in the script, our chance to massage verbiage occurs as we fill the T-chart, not after. The students need to see, as much as possible, that it is their ideas and their language that have contributed to the formation of the rubric. This can be heightened if, during the next class, they can see the new rubric and the original T-chart at the same time. So, if you can leave it on the board, great. Otherwise, show them a picture of the T-chart when you present them with the rubric that they had a hand in cocreating.

As the rubric will, in essence, be a series of indicators each of which is on a continuum, it is also important to select language from the T-chart that is dichotomous in nature. As you can see in Figures 12.2 and 12.3, each *good* indicator on the right is paired with an opposing *bad* indicator on the left (*open to other's ideas—closed to other's ideas*). Students are not necessarily thinking about this when they are offering suggestions for what goes into the T-chart. So, you may need to suggest that they think about what the opposite *bad* is for each statement in the *good* column. This is not to say that each suggestion will have a natural opposite. What is important is that the indicators that make it into the final rubric are dichotomous.

Using the Rubric

How you present the rubric, and how you use it, turned out to be just as important as how it was created. For Grades 4–12 we found it worked best if it could look something like this:

> **Teacher** So, yesterday you helped me create this T-chart. After school I took that chart and made it into a rubric. [Teacher projects the rubric.] You will see that one of these rubrics is taped up at each vertical surface. This is because today, while you are working at the boards, I will be evaluating three groups using this rubric.

Why only three groups? The answer to this is simple—you only have time to do three groups. Keep in mind, you are still having to manage flow and prepare for consolidation. It is important that you do not compromise your role as the teacher in the thinking classroom just to be an evaluator. This is still a thinking space that needs your stewardship.

For younger grades (K–3) we found that the introduction needed to be a little bit different.

> **Teacher** So, yesterday you helped me create this T-chart. After school I took that chart and made it into a rubric. [Teacher projects the rubric.] Let's go over it together so we all understand what each part means. [. . .] You will see that one of these rubrics is

taped up at each vertical surface. This is because today, while you are working at the boards, I expect all of you to behave like students in these pictures [teacher points at the right column].

Regardless, the perseverance in the room that day will be through the roof. And not because of the specter of evaluation, but because the coconstructed rubric makes it clear, to every student and every group, what is expected of them that day. The fact that you are evaluating three groups is only necessary to show that you are, truly, evaluating what you value.

> Once students see what behaviors are expected, and that these behaviors are valued, the students begin to see them as valuable as well.

At the end of the time spent on the boards, you give every group a highlighter and ask them to self-evaluate how well their group persevered that day. When they are done, give the three groups that you were evaluating (in Grades 4–12) the rubrics that you highlighted for them. The total time you will have spent on cocreating, presenting, and using this rubric will be less than 15 minutes—but the transformational changes in your students will be monumental. Every time we have done this, the changes in student behavior around whatever competency we choose to focus on, irrespective of how poor it was the day before, is huge. Once students see what behaviors are expected, and that these behaviors are valued, the students begin to see them as valuable as well. And when these competencies improve so does your thinking classroom.

FAQ

Q Should we be recording how each group did? If not, why would the students care—how would they see that we value it?

A This is good question. And it was something that we wondered as well. We are so used to using grades as the carrot or stick—depending on the student—that we use to motivate student behavior. What we learned from the research on rubrics is that students see that we value something if we are willing to spend grades on it—or if we are willing to spend time on it. They know that time is limited and, as such, it is valuable. And if you are willing to give some of it up to focus on their perseverance—or collaboration, or risk taking, or whatever behavior you want to focus on—then it must be valuable. This is not to say that you can't record their performance in some way. What is important, however, is that you do not let them see how you do it. Our research clearly showed that if you put a number or a letter on a rubric, then the students ignore all other aspects of the feedback that the rubric affords. Race (2010) as well as others have observed the same phenomenon across a variety of settings. Numbers and letters go in your grade book—not on the rubric.

> Students see that we value something if we are willing to spend grades on it—or if we are willing to spend time on it.

Q If it is true that the students see something as valuable if we spend time on it, why do we even have to bother saying we will evaluate three groups?

A What we learned is that students in Grades 4–12 are often so used to value being projected through the collection of grades that they, themselves, don't know any other way to gauge value. This is more prevalent in high school students, but we saw it to some degree in students as young as Grade 4. When you state that you are evaluating three groups, they see that you are valuing these groups' behavior through grades. When they self-evaluate their performance and see how valuable this feedback is, then they begin to see that there is value inherent in the process and the time spent on it. So, after two or three uses of rubrics in your classroom, you can drop the pretense of evaluating three groups.

> When they self-evaluate their performance and see how valuable this feedback is, then they begin to see that there is value inherent in the process and the time spent on it.

Q How often should we be cocreating and using rubrics?

A You cocreate a rubric whenever there is a behavior that you would like to improve within the room. Once you cocreate it, you need to use it right away and for two or three lessons in a row to really show that you value it. Then you can take a break from it for a while and only use the rubric once in a while, or when you see that it is needed. For example, if you see a particular group not persevering, just grab a perseverance rubric and tape it up on their vertical surface. This will signal to that group that you are not impressed with their behavior and that you will be watching them while at the same time communicating exactly what it is that you will be watching for. And once in a while you will see a group member pulling out an old rubric for their group because that individual, or the group as a whole, is aware that they aren't behaving as they should.

Q You mentioned that these new rubrics should only evaluate one competency at a time, but you didn't specify how many indicators within the rubric to look for. The example you gave in Figures 12.2 and 12.3 has five. And then the example of the T-chart in Figure 12.6 looks like it can end up with many more. Is there an optimal number of indicators that the rubric should have?

A We found that five was the maximum you should have. Beyond that, the rubric either starts to get redundant, or it begins to lose focus. It's true that the T-chart in Figure 12.6 will eventually have many more than five indicators. This means that you are going to have to be selective as you choose which are the most important to include in the rubric. Pick the ones most on point that are also most clear to the students.

Q Is it really true that K–1 students are not ready to see these rubrics as a continuum and only see their behavior as fitting in one column or the other?

A Yes and no. From a developmental perspective, nuance and subtlety are just beginning to emerge within this age group. But how fast it develops depends a lot on what kinds of educational experiences and how much exposure to the two-column rubrics they have had. We saw Grade 1 students who, by the end of the school year, were able to start using the rubric in Figure 12.3 as a continuum.

Q You talked about these coconstructed rubrics as *observational*— to be used only for evaluating students' visible in-the-moment behaviors. Can I also use them to evaluate students' *producibles*— things that they hand in?

A Yes—but the process of creating them is slightly different. One of the things we learned from the research is that, regardless of how poor a class may be around an observable competency such as perseverance or collaboration, they can always tell us what it looks like to be good at it. The same is not true for competencies that are evaluated through *producibles*. For example, if we gather students at the board and ask them to tell us what a good problem-solving solution looks like, or a good proof, they cannot tell us—at least not with the same confidence as when we ask about an observational behavior. You need to begin this process by having them look at *exemplars.* You give each group the same three exemplars of, for example, a problem-solving solution. One of these examples is very poor, one is very good, and one is in the middle. You ask them, as a group, to put these exemplars in order from worst to best and then discuss among themselves what it is that makes the good one good and the poor one poor. Then, when you gather the students at the board to coconstruct the T-chart, they have things in mind that they can offer. A small nuance that is important when using exemplars is to, as much as possible, make the middle exemplar the longest exemplar. This will push the students to discuss quality over quantity.

Q OK—competencies and *producibles* aside, how will this help me evaluate my students' attainment of content?

A It won't. This chapter, and the rubrics discussed here, are an answer to the question of how we show, through evaluation, that we value competencies like collaboration, perseverance, risk taking, and so on. This is not to say that these rubrics cannot be adapted to evaluate attainment of content—we just did not pursue that. Having said that, Chapter 13 looks at how we can help students self-assess how they are doing with the attainment of content, and Chapter 14 looks at how we can grade students on their attainment of content—while at the same time increasing student thinking and engagement.

Q You talk about these rubrics as evaluations. Shouldn't they be called assessments?

A In the assessment and evaluation literature, assessment is often defined as being formative, and evaluation is defined as being summative. You will see in the next two chapters that this dichotomy eventually collapses in on itself. All assessment and evaluation should be formative; some of it will also be summative. Likewise, I do not find the distinction between assessment or evaluation *of, for,* or *as* learning to be of much help. All assessment and evaluation should be *for* learning, and some of it will be *of* learning. And all assessment and evaluation should be *as* learning—as it was in this chapter with the coconstruction of the rubric. Taken together, I do not distinguish between assessment and evaluation, and I use the terms interchangeably. In this chapter I used *evaluation* because of its close connection to *values* and the prompting question around what it is we value in our students.

SUMMARY

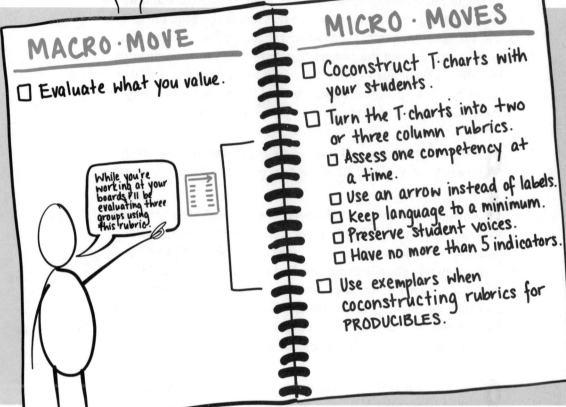

MACRO·MOVE

☐ Evaluate what you value.

While you're working at your boards I'll be evaluating three groups using this rubric.

MICRO·MOVES

☐ Coconstruct T·charts with your students.
☐ Turn the T·charts into two or three column rubrics.
☐ Assess one competency at a time.
☐ Use an arrow instead of labels.
☐ Keep language to a minimum.
☐ Preserve student voices.
☐ Have no more than 5 indicators.

☐ Use exemplars when coconstructing rubrics for PRODUCIBLES.

QUESTIONS TO THINK ABOUT

1. What are some of the things in this chapter that immediately feel correct?

2. If you have previously used a rubric that has more than three columns, take a good look at it, and see if you see language that is ambiguous and not helpful for moving students from one column to the next. If that rubric were scrambled, do you think you could unscramble it?

3. Think about some competencies that you feel your students need to improve on. Which of these do you think you should coconstruct a rubric for first?

4. In this chapter I mentioned that it is easiest to coconstruct a rubric right after an experience in which the class was deficient in the particular competency you want to focus on. With this in mind, think of some experiences that you can manufacture that will accentuate the deficiency you want to address first. For example, if you want to focus on perseverance, you can begin by giving them a task that is tempting to give up on, but is solvable with time and effort.

5. In the FAQ I mentioned that it is possible to coconstruct rubrics for *producibles*. What kind of producibles do you use that you would like your students to get better at? What would the exemplars look like?

6. What are some of the challenges you anticipate you will experience in implementing the strategies suggested in this chapter? What are some of the ways to overcome these?

☑ TRY THIS

The following tasks have been selected because they require a lot of perseverance on the part of the students and, therefore, would be a perfect precursor to coconstructing a perseverance rubric.

Grades K–3: How many 7s?

If I were to write the numbers from 1 to 100, how many times would I use the digit 7? What if I wrote 1 to 1,000? How many times would I use the digit 0?

$$1, 2, 3, 4, 5, 6, 7, 8, \ldots, 97, 98, 99, 100$$

Grades 4–9: Country road

A country road is 27 miles long and goes all the way around a lake, connecting the six cottages that are next to the lake. Two of the cottages are 1 mile apart (along the road). Two cottages are 2 miles apart, two are 3 miles apart, two are 4 miles apart, …, two are 25 miles apart, and two are 26 miles apart. How are the cottages distributed along the road? Find a second way to distribute them.

Grades 10–12: Pirate diamond

A band of nine pirates is going to disband. They have divided up all of their gold, but there remains one *giant* diamond that cannot be divided. To decide who gets it, the captain puts all of the pirates (including himself) in a circle. Then he points at one person to begin. This person steps out of the circle, takes his gold, and leaves. The person on his left stays in the circle, but the next person steps out. This continues with every second pirate leaving until there is only one left. Who should the captain point at if he wants to make sure he gets to keep the diamond for himself? What if there were 10 pirates? 11 pirates? N pirates?

Source: The pirate diamond task is an adaption of the "Josephus problem."

CHAPTER 13

HOW WE USE FORMATIVE ASSESSMENT IN A THINKING CLASSROOM

• •

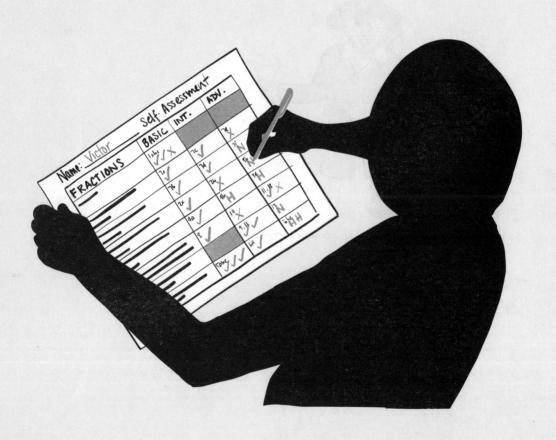

As you read in the previous chapter, the assessment and evaluation literature often draws a distinct line between assessment and evaluation. It generally defines assessment as being formative while it defines evaluation as being summative. I argue that this distinction is not really helpful, because in reality all assessment and all evaluation should be formative, and some of it will also be summative. This is why you'll see the terms *assessment* and *evaluation* used interchangeably within this book.

In the previous chapter we looked at the assessment of *competencies* and how the clear communication of competencies as *valuable* (through evaluation) can profoundly affect students' behaviors in the thinking classroom. In this chapter, and the next, we will look at the assessment of *content* and how, again, clear communication can have a profound impact on student behaviors. While Chapter 14 will focus on *summative* assessment of content, this chapter focuses on *formative* assessment. By the end of this chapter you will learn that if we, as teachers, are careful about what and how we communicate with our students, then we will see not only significant improvements in students' abilities to think about their own learning, but also significant improvements in students' attainment of content.

Formative assessment: is the gathering of information for the purposes of informing teaching and learning.

Summative evaluation: is the gathering of information for the purposes of grading and reporting.

 # THE ISSUE

Whether formative or summative, assessment is fundamentally about the communication of information. For much of the 20th century, this information was seen, almost exclusively, as flowing from the student to the teacher for the dual purposes of informing *teaching* and producing a *grade*. The methods used to collect this information were either formal (tests, quizzes, assignments, projects, presentations, portfolios, etc.) or informal (observation, conversations, etc.). Whereas grading was typically informed by the more formal means of assessment, teaching was informed by both formal and informal means.

For much of the 20th century, assessment was seen, almost exclusively, as the flow of information from student to teacher for the dual purposes of informing *teaching* and producing a *grade*.

In the last 20 years, however, there has been increasing attention paid to assessment practices wherein the flow of information is reversed, flowing from the teacher to the student, for the purpose of informing *learning*. Like the information that informs teaching and/

> In the last 20 years there has been increasing attention paid to assessment practices wherein the flow of information is reversed, transmitting from the teacher to the student, for the purpose of informing *learning*.

or grading, that which informs learning is communicated both formally and informally, often relying on the same methods mentioned above.

For example, when a student completes a test or a quiz and the teacher grades it, how that student performed informs the teacher of what they need to work on with that student. When that graded test or quiz is returned, it likewise informs the student what it is they need to work on. Similarly, if a teacher has a conversation with a student, that teacher is collecting information about what that student does or does not understand and what they can or cannot do. If, during this conversation, the teacher provides feedback about what they gleaned from the interaction, then the student is likewise being informed about their own understanding. In either case, the teacher can then use the information to inform their *teaching*, and the student can use it to inform their *learning*.

THE PROBLEM

The problem is that, although the information that is flowing from the student to the teacher is comparable to the information that is flowing from the teacher to the student, the recipients of the information and what they can make of it vary greatly. The teacher is interpreting the information against a background of full knowledge and understanding of the concept or concepts in question, a clear picture of what comes next, and a rich history of having taught this same concept many times. The student, on the other hand, has an incomplete—and maybe inaccurate—picture of the concept, has no idea what it is leading to, and is learning it for the first time. How could the same information possibly inform them in the same way?

> Although the information that is flowing to the student is the same as that flowing to the teacher, the recipients of the information and what they can make of it vary greatly.

I'll give a very simple example of this. I interviewed students in many different mathematics classes at many different grade levels just as they were finishing up a unit of study. I asked just one question.

I have never asked a question that is so predictive of student performance on a unit test. About 15% of the students told me that the unit they just finished was made up of a number of subtopics, and they were able to name or describe what those subtopics were. These students, for the most part, scored above 90% on the upcoming test. There was a group that could tell me that there were subtopics, but they were not able to completely delineate them or describe them. These students tended to score between 75% and 90%. The rest of the students all said that the unit was just one big topic. These students tended to score below 75%.

How can a student in this last category—who, for example, sees subtraction of two-digit numbers as one big topic—possibly make sense of feedback that they still need to work on subtracting two-digit numbers where decomposition is needed? Contrast this with how clearly you, as a teacher, view a unit of study as a collection of subtopics, sections, and/or special cases and how this allows you to clearly see that this is exactly what this student needs to work on. Put in a different way, information communicated from a teacher to a student who sees the topic as one big unit will only inform that student of *what it is that they can do*; but because they don't have a clear picture of the whole unit and all its subtopics, they cannot see what there is still left to learn. The teacher, on the other hand, with their greater sense of the scope and scale of the topic, can use the information that is communicated from the student to determine *what that student can do* AND *what they cannot yet do*. In order for assessment to equally inform teaching *and* learning, we need to find ways to help students see mathematical topics as collections of subtopics, sections, and/or special cases the way teachers do, and to use this knowledge to inform themselves about what it is they can and cannot yet do.

> Information communicated from the teacher to a student, in many cases, can only inform that student of *what it is that they can do*, but the teacher can use the information that is communicated from the student to determine *what that student can* AND *cannot yet do*.

TOWARD A THINKING CLASSROOM

For someone to be able to navigate, by land or by sea, they need two pieces of information—where they are and where they are going. Both pieces of information are vital and of equal value. If they don't know where they are going, they are destined to get lost. And if they don't know where they are, well, then they are already lost. The same is true of students trying to navigate their own learning—they need to know where they are and where they are going. In the context of a thinking classroom—or any classroom—where they are is what they understand, know, and/or are able to do. And where they are going, within the scope of a unit of study, is what they have not yet learned, don't yet understand, and/or are not yet able to do. Frey, Hattie, and Fisher (2018) refer to students who can navigate their learning in these ways as "assessment capable visible learners."

> To help students navigate their learning, the information we need to communicate to them is the information that helps them know not only what they know, but also what they don't know.

To help students navigate their learning, then, the information we need to transmit to them is the information that helps them know not only what they know, but also what they don't know. And to do that, we need to first help them to see what the subtopics are that make up the background against which we assess what they know and what they don't know. To achieve this, we experimented with feedback on tests, quizzes, and check-your-understanding questions. What we found in the end, however, was that the nature of the feedback that we provided was important, but not as important as how we help students to retain and organize the feedback. This, coupled with the idea of trying to help students know where they are and where they are going, allowed us to eventually converge on the creation and use of instruments like the ones in Figures 13.1, 13.2, and 13.3.

REPEATING PATTERNS	BASIC	INTERMEDIATE	ADVANCED
Identify the core of a repeating pattern	ABABAB... abbcabbcabbc... ◇◇□◇◇□◇◇□...	ABABBCABABBC... O□◇◇O□◇◇...	
Transfer a repeating pattern	ABAB...=O□O□... ffyh...= O□O◇O□O◇...=		
Extend a repeating pattern	ABABAB▲⊠▲ nnjknnjk___ O◇O◇O◇___	$+$+!$+$+!___ KLMNOKL___	
Fill in the blanks in a repeating pattern	ABAB AB XYZXYZ _YZ... □□O◇_□O◇...	abccda_ccd... _#$$H#$$...	A_BCAB_C... _O◇◇_OO_O□...
Create a repeating pattern	_____...	_____...	

Figure 13.1 Instrument for navigating where you are and where you are going for repeating patterns.

ADDITIVE NUMBER PATTERNS	BASIC	INTERMEDIATE	ADVANCED
Identify the rule for a number pattern	2,4,6,...[+2] 1,3,5,... 4,8,12,...	1,6,11,... 7,10,13,...	23,30,37,... 17,28,39,...
Extend a number pattern	2,4,6,8,10,12 3,6,9,_,_,_ 6,12,18,_,_,_	1,5,9,_,_,_ 9,16,23,_,_,_	22,31,40,_,_,_ 41,63,85,_,_,_
Fill in the blanks in a number pattern	2,4,6,8,10,... 5,_,15,20,... 4,8,_,16,...	1,4,_,10,... 8,_,18,23,...	_,16,22,... 3,_,17,...
Create a number pattern	_,_,_,_,_,...	_,_,_,_,_,...	_,_,_,_,_,...

Figure 13.2 Instrument for navigating where you are and where you are going for additive number patterns.

FRACTIONS	BASIC	INTERMEDIATE	ADVANCED
Definitions	1abc		
Add and subtract proper fractions	7a	7c	7e
Add and subtract mixed fractions	7b	7d	7f
Multiply and divide proper fractions	2a	2b	3b
Multiply and divide mixed fractions	4a	4b	3c
Solve order of operation tasks with proper and mixed fractions	8	10	11, 13
Solve contextual problems involving fractions		9, 11	12
Estimate solutions for problems involving fractions	5abc	6a	6bc

Figure 13.3 Instrument for navigating where you are and where you are going for fractions.

Being explicit about the list of outcomes that constitute a unit of study helps students understand that a topic is comprised of a collection of subtopics.

These navigation instruments are created by the teacher and are, essentially, a way for students to self-evaluate their performance on an individual quiz, review test, or set of check-your-understanding questions. In each example, the leftmost column is a list of subtopics for a specific unit of study and constitutes the list of outcomes within that unit. These can come from the curriculum, the textbook, or from a teacher's own understanding of what needs to be learned. What we learned was that being explicit about the list of outcomes that constitute a unit of study helps students understand that, for example, fractions is not one big topic, but a collection of subtopics.

Each of these subtopics (or outcomes) can be further broken down by conceptual complexity (basic, intermediate, and advanced). Take for example, this list of adding fraction questions:

1. $\frac{1}{5} + \frac{3}{5}$

2. $\frac{1}{4} + \frac{3}{8}$

3. $\frac{3}{5} + \frac{1}{7}$

You will immediately recognize that this is not a random list of questions. Although these all require a student to add two fractions, there is a marked difference in the skills, knowledge, and understandings needed to answer each question. How a student performs on each of these provides distinct information regarding whether their knowledge, skills, and abilities are at the *basic*, *intermediate*, or *advanced* complexity level. Can they add fractions with the same denominator (basic)? Can they add fractions where the denominators are different, but one is a multiple of the other (intermediate)? Can they add fractions where the denominators share no common factors (advanced)? Which of these questions they can successfully answer communicates different information about their level of mastery of the outcome *adding fractions* and, thus, needs to be differentiated from each other. The three rightmost columns in the navigation instruments (Figures 13.1, 13.2, 13.3) do just that.

Once this table has been created, the teacher populates it with questions (Figures 13.1 and 13.2) or question numbers (Figure 13.3) from the quiz, review test, or check-your-understanding questions such that each question in the table is situated within the correct outcome and complexity level. Note that some outcomes do not have all the complexity levels. For example, *transfer a repeating pattern* (Figure 13.1) only exists at the basic level. Conversely, *solve contextual problems involving fractions* (Figure 13.3) does not even have a basic level. These omissions are represented by blocking out the irrelevant cells with shading, as in Figures 13.1 and 13.3. Likewise, if a question requires students to draw on multiple outcomes to solve it, such as Question 11 in Figure 13.3, then it sits in more than one place within the table.

This linking of a specific question to an outcome turned out to be vital. Although the language in the left-hand column is clear to us, students needed to see specific questions to fully understand what many of the outcomes meant. "Factoring a quadratic where the leading coefficient is equal to one" doesn't mean a lot to many students until they see that this outcome is composed of questions that look like this:

- $x^2 + 5x + 6$

- $x^2 + 14x + 24$

- $x^2 - x + 12$

- $x^2 - 6x - 16$

- …

Likewise, "adding two digit numbers with regrouping" may not make a lot of sense until they see that this is referring to questions that look like this:

- 23 + 58

- 67 + 71

- 39 + 86

- 48 + 62

- . . .

This turned out to be especially true in the primary grades, where students' reading abilities are just beginning to emerge—and, even then, not for technical mathematical language. Depending on the grade, the teacher may wish to create navigation instruments without the leftmost column altogether (see Figures 13.4 and 13.5). Without the outcome to tell students what is expected, clarity must be communicated differently. In the following examples, this is done through the inclusion of an example for each outcome at the basic complexity level.

BASIC	INTERMEDIATE	ADVANCED
<u>AB</u>ABAB... abbcabbcabbc... ◇○□◇○□◇○□...	ABABBCABABBC... ○□○◇○□○◇...	
ABAB...=○□○□... ffyh...= ○□○◇○□○◇...=		
ABABAB▲🅱▲ nnjknnjk___ ○◇○○◇○___	$+$+!$+$+!___ KLMNOKL___	
ABA🅱AB XYZXYZ_YZ... □○○◇_□○◇...	abccda_ccd... _#$$H#$$...	A_BCAB_C... _○◇○_○○_◇□...
_ _ _ _ _ _ _ _ _ ...	_ _ _ _ _ _ _ _ _ ...	

Figure 13.4 Navigation instrument without the outcomes (repeating patterns).

BASIC	INTERMEDIATE	ADVANCED
2,4,6,...[+2] 1,3,5,... 4,8,12,...	1,6,11,... 7,10,13,...	23,30,37,... 17,28,39,...
2,4,6,8,10,12 3,6,9,—,—,— 6,12,18,—,—,—	1,5,9,—,—,— 9,16,23,—,—,—	22,31,40,—,—,— 41,63,85,—,—,—
2,4,6,8,10,... 5,—,15,20,... 4,8,—,16,...	1,4,—,10,... 8,—,18,23,....	—,16,22,... 3,—,17,...
—,—,—,—,—,...	—,—,—,—,—,...	—,—,—,—,—,...

Figure 13.5 Navigation instrument without the outcomes (additive number patterns).

Regardless of grade, linking specific questions to different complexity levels (basic, intermediate, and advanced) further helps students to more completely understand where their learning is with respect to a specific topic. For example (see Figures 13.2 and 13.5), being able to extend the pattern 1, 5, 9, ... (intermediate) is not the same as being able to extend the pattern 41, 63, 85 ... (advanced).

As mentioned, these navigation instruments can be used by a student to self-evaluate how they performed on a quiz, review test, or set of check-your-understanding questions. Regardless of how it is used, however, performance on a specific question goes beyond whether it is done correctly (✓) or incorrectly (✗). In order for the navigation instrument to work properly, students need to also distinguish between whether they completed a question with or without help, whether they got a question correct or if they made a silly mistake, and whether they got a question wrong as opposed to not trying it at all. Each of these cases carries unique information that is useful to track. To facilitate this nuanced tracking we used six symbols: ✓, S, H, G, ✗, and N.

> Students need to also distinguish between whether they completed a question with or without help, whether they got a question correct or if they made a silly mistake, and whether they got a question wrong as opposed to not trying it at all.

✓ questions that are attempted and answered correctly

∫ questions that are attempted and mostly answered correctly, but have a silly mistake

H questions that are attempted and answered correctly with help from the teacher or a peer

G questions that are answered correctly within a collaborative group

X questions that are attempted and answered incorrectly

N questions not attempted

For example, the clear demarcation of learning outcome and complexity level in Figure 13.3, coupled with the student's nuanced records of how they performed on each question (see Figure 13.6) helps them to clearly see what they are able to do (where they are) and what they are not yet able to do (where they are going).

FRACTIONS	BASIC	INTERMEDIATE	ADVANCED
Definitions	1abc ✓✓✗		
Add and subtract proper fractions	7a ✓	7c ✓	7e ✗
Add and subtract mixed fractions	7b ✓	7d ✓	7f N
Multiply and divide proper fractions	2a ✓	2b ✗	3b N
Multiply and divide mixed fractions	4a ✓	4b H	3c H
Solve order of operation tasks with proper and mixed fractions	8 ✓	10 ✗	11, 13 ✓ ✗
Solve contextual problems involving fractions		9, 11 ✓ ✓	12 N
Estimate solutions for problems involving fractions	5abc ✓✓✓	6a ✓	6bc H H

Figure 13.6   Student's record of how they did on the fractions practice test.

In our initial trials we used this navigation instrument as a way for students to record how they had performed on an end-of-unit review test—with astonishing results. When these students then wrote their end-of-unit test, we saw an immediate improvement in grades of 10%–15% for 50%–70% of the students. For many of these students, the knowledge of where they were and where they were going was all they needed to help them improve.

> For many of these students, the knowledge of where they were and where they were going was all they needed to help them improve.

Jamal I mean, now I know exactly what I need to work on.

And some students expressed that they finally understand how a unit of study is broken into subtopics and what those subtopics are.

Angel I finally get what we are doing.

Colleen Are you kidding me? This is great. I know what we are doing now.

The question is, why do we see improvements in only 50%–70% of the students? Well, some of your students already see the subtopics in what they are learning. And, for the most part, these students already know where they are in their learning of these subtopics. For them, this navigation instrument provides redundant information and, thus, produces no improvements. But this only accounts for 10%–20% of your students.

Another group on whom this navigation instrument was having very little impact were the students who didn't appear to care about either their learning or their grade. Some of these students were performing at the lower end of their respective classes. For these students, information about where they are and where they are going wasn't helping them to move forward. They already knew where they were, and they didn't really have ambitions to go anywhere else. This is not to say that they couldn't be helped. Just not in this way. This is also not to say that all students performing at the lower end of their class didn't care. We saw big improvements in many students who were previously performing poorly—many moving from failing to passing a unit test. For these students, the unit of study was a complex collage of things they didn't know how to do. The navigation instrument cut through all that noise and gave them a map that allowed them to focus only on the easy questions—something they could see as achievable.

However, when we accounted for the students who already saw the subtopics and the students who didn't care, there were still students who did not improve. These students were mostly achieving B's, and even the teacher had difficulty figuring out what bound this demographic together. It took interviews with these students to begin to untangle what was happening.

Jordy　　Hey, I got a B . . . without doing ANYTHING. Why would I want to put in a bunch of work to try to get an A?

Steph　　A B is good enough for my mom.

Chris　　I'm not one to go the extra mile.

It is not that this group doesn't care. They do. But what they care about are grades that are "good enough" and what they don't care about is "going the extra mile." This group of students got to where they were with little effort and, for them, any more work would produce diminishing returns. From the perspective of economy of effort, these students are not wrong. If you can achieve your goal with zero effort, no improvement would be worth more effort. This way of thinking is difficult to argue with. These students care only about the grade—which brings us back to the idea of evaluating what we value (Chapter 12).

Who is left are the students who care about their learning and want to improve (50%–70%). For these students, a navigation instrument like the one in Figures 13.1, 13.2, 13.4, and 13.5 not only provides information about where they are and where they are going, but also does so in a way that is clear to them. And with this clarity they can begin to navigate their learning—to think about their learning.

Q Is this the only way for students to understand where they are and where they are going?

A No. In fact, the rubric in Chapter 12 achieves this as well. Where the rubric is highlighted is where the students are on the continuum, and the descriptors in the right hand column is where they are going. A clothesline, often used in elementary school to display student work on a continuum, is another example of a way to help students understand where they are and where they are going.

Q I give students lots of written feedback on their quizzes and tests to help them understand where they are in their learning. Isn't that enough?

A For the students who see the unit of study as subtopics, it is. They already have a very clear picture of where they are going, and your feedback is helping them understand where they are. For everyone else, your feedback is not enough. In our research into this topic, we found that, in general, teachers are good at providing feedback either about where students are or where they are going. Very few provide both. At least not in ways that are clear to the students.

Q For my whole career I have just given back quizzes and tests with each question graded. From this the students can see what they got right and what they got wrong. And for the ones they got wrong, what parts they got wrong and what grade they got on that question. Are you saying this is not helpful?

A This is an example of what I call *encrypted feedback*. You, and the students who see all the subtopics, have the decryption key and can make sense of the feedback. For everyone else, however, all they see is scores and whether the score is good enough or not.

Q In my jurisdiction we wouldn't use the headings of *basic, intermediate*, and *advanced*. We would instead use *novice, emergent*, and *expert*. Does that make a difference?

A Yes, it makes a big difference. First, our research initially showed that *easy, medium*, and *hard* were the headings that the students found the clearest. And this whole chapter is about communicating with students in a clear way. After reading Tracy Zager's (2017) book, we switched the headings to *basic, intermediate*, and *advanced* with no degradation of student clarity or preference. The advantage with *basic, intermediate*, and *advanced* was that teachers preferred these headings over *easy, medium*, and *hard*.

Second, whereas *basic, intermediate*, and *advanced* (and *easy, medium*, and *hard*) identify the complexity level of

> Whereas headings like *basic, intermediate*, and *advanced* (and *easy, medium*, and *hard*) identify the complexity level of the questions, headings like *novice, emergent*, and *expert* describe the abilities of students. This makes a big difference.

the questions, headings like *novice*, *emergent*, and *expert* describe the abilities of students. This makes a big difference. You can label the headings any way you want, as long as you are talking about the complexity level of the questions or concepts, not the abilities of the students.

Q In Chapter 12 you dispensed with heading altogether and opted instead for an arrow. Why not do the same here?

A We thought the same thing. So, we tried this in early iterations of the research into helping students see where they are and where they are going. It worked well to a point, but once students and teachers began to talk about the feedback and the complexity level of different outcomes, they spontaneously began to create names for the different columns (*first, second, third; one, two, three*; etc.). Once we realized that headings were inevitable, we began experimenting with different headings with the end result being what is in the navigation instrument in Figures 13.1, 13.2, 13.4, and 13.5.

Q I can think of a few of my students who would see the headings *basic*, *intermediate*, and *advanced* and just opt to do only the *basic* questions. Isn't that a problem?

A Before I answer this question I want to differentiate between the students that would use these headings as a way to do less work and less thinking and the students who would use these headings as a way to find an entry point into the learning of the topic. This question is about the former type of students. For these students this is a problem. But the problem is with the students, not with the navigation instrument. And, for this reason, the solution lies not in the instrument, but within the students. The real question is not how to change the instrument, but how to get students who don't care about learning to care about learning. I won't address this here, but I can say that for students who used these headings as an entry point into learning the topic, the heading offered them a place to start (*basic*). Once they clearly saw where they were and what came next (*intermediate*), they were more likely to want to take the next step—they wanted to level up.

> For students who used these headings as an entry point into learning the topic, the heading offered them a place to start (*basic*). Once they clearly saw where they were and what came next (*intermediate*), they were more likely to want to take the next step—they wanted to level up.

Q I notice that the pattern navigation instruments (Figures 13.1, 13.2, 13.4, 13.5) have the actual questions in the cells, while the fraction navigation instrument (Figure 13.3) had question numbers. Is this because the pattern instrument is for primary grades, and the fraction instrument is for intermediate grades?

A Not at all. Although we found that putting actual questions into the navigation instrument worked well for students of all grades, the length of some of the questions at higher grades made this prohibitive. However, using question numbers did not work well for primary students. The extra step of mapping a question number to a question and back to an outcome/complexity level was too much mapping. At the same time, a list of question numbers did not allow primary students to discern what bound a set of questions together to make an outcome.

Q In this chapter you mention that the teacher would populate the navigation instrument with the question numbers from the quiz, review test, or check-your-understanding questions. Wouldn't it be better if the students were able to decide for themselves what outcome (or subtopic) and what complexity level a specific question was?

A Yes. But our research showed that only about 10%–20% of students were able to do this right away. The rest needed several experiences with the teacher-populated navigation instrument before they were able to clearly identify a question as belonging to a specific subtopic and complexity level. The goal, of course, is that students get to this point. Just be ready for it to take most of the school year to achieve it.

Likewise, the goal is that the students begin to be able to identify and name subtopics on their own as they encounter them in a unit of study. This is especially important for students in Grade 12 and intending to go on to some form of postsecondary education. For them, being able to see subtopics in whatever they are studying will be a distinct benefit. But, again, do not expect that they can do this just because they are in Grade 12. However, we found that after repeated experiences with the navigation instrument, most students were able to disaggregate a unit of study into its relevant subtopics by the end of the school year.

Q I have a problem with all this splitting up of a unit of study—first into subtopics and then into different complexity levels. Don't we want students to see mathematics as connected? This just feels like it is disconnecting and compartmentalizing mathematics.

> **For students to see mathematics topics (or subtopics) as connected, they must first see them as distinct.**

A This is a very good question. We were concerned about this as well. But an interesting observation emerged from our research into the use of these types of navigation instruments to help students see where they are and where they are going. It turns out that for students to see mathematics topics (or subtopics) as connected, they must first see them as distinct. For example, multiplying mixed fractions (*advanced*) is not the exact same thing as multiplying proper fractions (*basic*). There are parts of these subtopics that are similar—cancelling. But there are parts that are different—turning mixed fractions into improper fractions. What allows these subtopics to be seen as connected is recognizing what parts are similar while at the same time being cognizant of the parts that are different.

For the students who saw units of study as one big topic, there were no connections. For them it was a bunch of disconnected and discrete routines to memorize. Once they could clearly see the different complexity levels, they could begin to see how, for example, multiplying proper fractions is a special (and *basic*) case of multiplying mixed fractions (*advanced*). They needed to see the distinction to see the connections.

Q In my jurisdiction we have been told that we should be stating the learning goal, or outcome, for that lesson at the beginning of the lesson. Doesn't that take care of helping students see a unit of study as a collection of subtopics?

A In theory, yes. In reality, however, it doesn't. Many of the students in our research who could not identify a unit of study as consisting of subtopics were in classes where the learning goal was stated at the beginning of every lesson. In part, this is because a learning goal stated prior to a learning experience makes very little sense to students. For example, saying to a Grade 10 student that "today we are going to learn how to factor trinomials where the leading coefficient is not one" makes as much sense as me saying to you that "in the next chapter we are going to learn how to find the minimum spanning tree of an edge-weighted undirected graph." Until your students get to experience the mathematics and see how the different tasks in this subtopic are connected *and* are different from other tasks, these kinds of statements aren't going to mean much. Names of concepts should come after experiences with concepts. This is, in part, why consolidation from

> **Names of concepts should come after experiences with concepts.**

the bottom (Chapter 10) is so effective—it names ideas after students have experienced them.

By the way, in the next chapter we are not going to learn how to find the minimum spanning tree of an edge-weighted undirected graph. We are going to learn about the difference between the point-gathering and data-gathering paradigms and how these paradigms have bifurcated our ideas around aggregating and grading.

Q Ultimately, this chapter is about self-assessment—you actually mention this in several places. How is this different from the many other forms of self-assessment that have been used for the last several decades?

A Self-assessment, as used in classrooms for the last few decades, has largely been based on students' *opinions* of their abilities. The navigation instrument, on the other hand, is a form of self-assessment that is based on *data* about students' abilities. This, it turns out, makes a big difference not only about what self-assessment captures, but also how students use and perceive the feedback from the self-assessment. We found that students, for the most part, took the data coming out of the navigation instrument very seriously. The same was not true of the feedback coming out of opinion-based self-assessments.

> Self-assessment, as used in classrooms for the last few decades, has largely been based on students' *opinions* of their abilities. The navigation instrument, on the other hand, is a form of self-assessment that is based on *data* about students' abilities.

In truth, it could be argued that opinion-based self-assessment is not a form of feedback at all. One student in our research referred to these types of self-assessments as "feedforward," because students tell it what they can do, not the other way around.

Q Isn't the use of this type of navigation instrument not just a type of outcomes-based assessment?

A Yes and no. It is a form of outcomes-based assessment in that it is delineating outcomes for the purpose of assessment. But outcomes-based assessment is typically used to describe how outcomes are used in grading and reporting. Although this navigation instrument can be used for that purpose, this chapter is focused on how to communicate to students where they are and where they are going vis-à-vis outcomes and complexity levels. In this regard, you might say that it is a form of outcomes-based self-assessment. How to use this instrument additionally for the purposes of grading and reporting will be discussed in the next chapter.

Q How can I know that I am doing a good job helping my students know where they are and where they are going?

A At the end of a unit of study, ask your students to make a review test on which they will get 100%. If they can do this, then they know what they know. Then ask them to make a review test on which they will get 50%. If they can do this, then they know what they know *and* they know what they don't know.

Q You introduce the navigation instrument as something that you used to have students self-assess how they did on a review test. But you also mention that it can be used to self-assess check-your-understanding questions. How are these different?

A When we first developed this navigation instrument, we used it exclusively to self-assess on the review test at the end of a unit of study. Such a use tells students what they can and cannot do and highlights which outcomes and complexity levels they have encountered that they have yet to demonstrate attainment of. When the navigation instrument is used to self-assess check-your-understanding questions, it does so throughout the whole unit of study. In addition to showing students their attainment of outcomes and complexity levels, this way of using the navigation instrument can also show growth over time. So, a student who could not do outcome X the first time they encountered it in a check-your-understanding question can see, over time, how they are now able to do *basic* questions, *intermediate* questions, and eventually *advanced* questions for this outcome.

Q At no point in this chapter do you mention that this navigation instrument can be used by students to record how they did on a test. Why not?

A Absolutely it can be used to record how they did on a test. I will speak to this more in the next chapter. From the perspective of communicating where you are and where you are going, however, recording how they did on a test is only useful if there is a retest. If there is no retest, and the test is the culminating experience for a unit of study, then there is nowhere else to go—so there is no need to keep navigating. On the other hand, if there is a retest or there are further opportunities for students to continue to demonstrate learning—which I highly support—then the journey is still on, and students need to continue to navigate.

Q I notice that you have three columns again—like in Chapter 12. Is this for the same reason?

A Yes. It turns out that almost any subtopic can be divided into three complexity levels in a clean and unambiguous fashion. For some we can go to four or five levels, but then we are constantly arguing about which level some questions fall into. Three levels were the clearest for the teachers and for the students.

Q How can I use the navigation instrument with my students who have modified or adapted learning plans?

A This navigation instrument is a great way to help students with modified or adapted learning plans to focus their, and your, attention. For example, you may have students for whom the learning plan is that they be able to do all the basic questions within a unit. Because the navigation instrument allows you to clearly identify these, the student has a clearer path forward.

Q Isn't it a lot of work to set up and use this type of navigation instrument?

A Not really. My experience is that if a teacher can sit down and create a unit test without referring to any resources, then they can create one of these navigation instruments off the top of their head. If it is the first time you are teaching a curriculum or you, yourself, do not see all the subtopics, then use your resources. Most resources are organized by subtopic and complexity level already—you just need to map them into the grid.

SUMMARY

MACRO · MOVE

☐ Help students to see where they are and where they are going.

> I mean, now I know exactly what I need to work on.

> I finally get what we are doing.

MICRO · MOVES

☐ Construct a navigation instrument like the one in figure 13.1.

☐ Use headings that delineate question complexity levels (not student abilities).

☐ Use the navigation instrument to help students record achievement on quizzes and review tests.

☐ Use the navigation instrument to help students record continuous progress on check · your · understanding questions.

QUESTIONS TO THINK ABOUT

1. What are some of the things in this chapter that immediately feel correct?

2. Which of your students see the subtopics within a unit and which do not?

3. Do you have some students who fall into that category where they are happy being good enough?

4. Can you think of ways in which you have previously received, or given, feedback that does not help a learner understand where they are and where they are going? If so, what information did the feedback communicate?

5. Can you think of ways in which your feedback has ever been encrypted in a way that obfuscates where students are and where they are going?

6. Can you think of other ways in which you can help students understand where they are and where they are going?

7. If an assessment instrument does not communicate where students are and where they are going, then who does that instrument serve?

8. What are some of the challenges you anticipate you will experience in implementing the strategies suggested in this chapter? What are some of the ways to overcome these?

☑ TRY THIS

Construct a navigation instrument like the one in Figure 13.1 or 13.3 for your next unit of study. You can lift the outcomes out of your curriculum or resource. Or, if you wish, you can decide for yourself what the outcomes should be. If you are choosing the latter option, there is a trick you can use to do it. Start by making an end-of-unit test. As you are making the test, every time you say to yourself, "I need one of these," write down in your own words what *one of these* is—for example, a growing pattern, adding two-digit numbers without regrouping, graphing a sinusoidal curve with a period other than 2π, et cetera. At the same time, whenever you find yourself saying, "I need two of these, one of these, and one of those," pause and ask yourself what differentiates *these* from *those* and how that places them at different complexity levels (*basic*, *intermediate*, and *advanced*).

CHAPTER 14

HOW WE GRADE IN A THINKING CLASSROOM

· ·

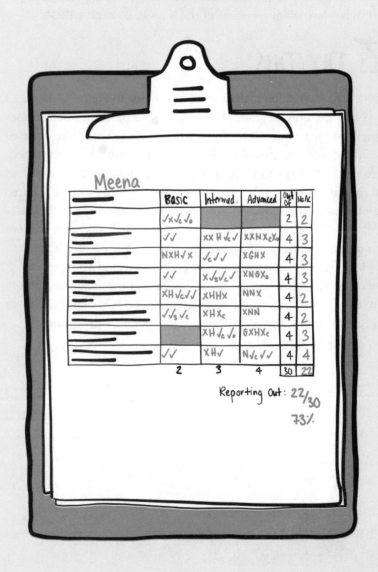

Meena

	Basic	Intermed.	Advanced	Out Of	No/c
	✓x ✓c ✓o			2	2
	✓✓	xx H ✓c ✓	XXNXcXo	4	3
	NXH✓x	✓c✓✓	XGHX	4	3
	✓✓	x✓s✓c✓	XNGXo	4	3
	XH✓c✓✓	XHHX	NNX	4	2
	✓✓s✓c	XHXc	XNN	4	2
		XH✓c✓o	GXHXc	4	3
	✓✓	XH✓	N✓c✓✓	4	4
	2	3	4	30	22

Reporting Out: 22/30

73%

As mentioned in the introduction, thinking classrooms emerged out of my efforts to break away from institutional norms and find new practices that not only occasion thinking, but also sustain thinking, particularly in—but not limited to—mathematics education. Constrained only by a set bell schedule and the four walls of a classroom, practices emerged that did just that. But no matter how much freedom we had to break the institutional norms within the classroom, three to four times a year we still had to dock with the mother ship and take all of the thinking and learning that was happening day-to-day in our thinking classrooms and report out a grade. In this chapter, you will learn how to do this in a way that not only honors the work that students are doing in a thinking classroom, but also continues to push your students to think about their own learning.

THE ISSUE

If you have implemented each of the thinking practices in Chapters 1–11, you will be feeling a tension between how you are teaching your students and how you are assessing them. Some of this tension should have been relieved by what was discussed in Chapter 12 (evaluate what you value) and Chapter 13 (formative assessment). But still, your students are spending so much of their class time learning in groups that individual tests may be beginning to feel disingenuous. At the same time, you may have students for whom your day-to-day subjective observation about what they are capable of does not align with how they perform on tests. If this is the case, then you are likely still feeling some tensions between how you are teaching and how you are grading.

Your students are likely also feeling these tensions. One of the most enduring institutional norms over the last 100 years is that learning days should, at least in part, resemble testing days. This is why there has been such a pervasive drive in the history of education to have students work individually for, at least, part of every lesson—it's a rehearsal for the test. Students know this. If learning days are now full of collaboration, and if learning days are rehearsal for tests, then why are tests still done individually? On many occasions in our research, students voiced exactly this question to the researchers. And in some cases, they asked it profusely and vociferously of their teachers.

> If learning days are now full of collaboration, and if learning days are rehearsal for tests, then why are tests still done individually?

To resolve these tensions, you may have been toying with the idea of changing the way your tests look, and you may be thinking about including collaboration in them in some shape or form. Alternatively, or additionally, you may also be toying with the idea of finding ways to grade some of the student learning you are seeing on a day-to-day basis in the thinking classroom.

THE PROBLEM

The problem is that when you get to the point where you want to make these changes in your grading practice, you face a new set of conundrums. If, for example, you move to using group quizzes or a group test, how do you distribute the grades that a group receives among its individual members? If one member of the group did nothing, or one member did all the work, is it then fair if everyone gets the same grade? Likewise, if you get to a point where you want to record and honor some of the day-to-day work that you observe in a thinking classroom, how would you record that in your gradebook, and how would you merge it with your test grades? Even if you work in a jurisdiction where it is either suggested or required that you triangulate each student's performance using observation, conversation, and product, these suggestions or mandates are often not accompanied by practical suggestions for how to implement them. Tensions beget tensions.

You are not wrong to feel this way. There is a real need to assess students collaboratively as well as through day-to-day observations in the thinking classroom. This need is supported by contemporary thinking on assessment and evaluation—see for example, O'Connor (2009) or Stiggins et al. (2006). And the tensions you are feeling exist, ultimately, because you are trying to make sense of how your 21st century thinking about grading can fit into a 20th century gradebook.

Grading practices over the last one hundred years can be seen as fitting into one of two paradigms—the point-gathering paradigm and the data-gathering paradigm.

The Point-Gathering Paradigm

The more enduring and more prevalent paradigm is what I call the *point-gathering* paradigm. I call it this because of the subtext that exists within the discourse in and around this practice.

Teacher	Don't forget that there is a quiz tomorrow.
Subtext	Tomorrow there are 20 points up for grabs—let's see how many you get.
Teacher	Today we are reviewing for Monday's test.
Subtext	Today I will be trying to help you get as many points as possible on Monday's test.
Teacher	This project is worth 20% of your final grade—so don't leave it to the last minute.
Subtext	I am giving you a chance here to get a LOT of points. Don't lose any points by leaving it to the last minute.

In the point-gathering paradigm, every point that a student manages to accrue is recorded in your gradebook, and at reporting time you take the number of points a student earned and divide it by the number of points they could have earned (with some scaling), and out pops a percentage. Within this paradigm, if a student receives a zero on anything, it affects the denominator but not the numerator in this calculation. And if a student gets any bonus grades, this affects the numerator but not the denominator. I also refer to the point-gathering paradigm as *event-based* grading because of the way these points are recorded in your gradebooks—with the name of the event (quiz, unit test, project, etc.) and the date on which the event occurred.

> There is a real need to assess students collaboratively as well as through day-to-day observations in the thinking classroom.

This was the dominant paradigm in the 20th century and, even now, is the most common practice used in mathematics classrooms in North America and many places around the world. Events-based grading is appealing and popular because it produces an objective grade that is believed to be an accurate reflection of what the student has learned. The problem is that, if the goal is to produce a grade that is reflective of what students have actually learned, then events-based grading is neither objective nor accurate.

> The problem is that, if the goal is to produce a grade that is reflective of what students have actually learned, then events-based grading is neither objective nor accurate.

The word objective means "expressing or dealing with facts or conditions as perceived without distortion by personal feelings, prejudices, or interpretations" (Merriam Webster, 2020). Lew Romagnano argues, in *The Myth of Objectivity* (2001), that because we have relationships with our students, built up over the year or years, we cannot avoid our appraisals being distorted by our *personal feelings, prejudices, or interpretations*. I'll give you an example of how you can know this is true.

Assuming that you are currently using, or have in the past used, some form of event-based grading, when you hit the equal sign and the percentages are calculated, do you look back across the row to see if the percentage for each student makes sense? I'm willing to bet you do. We all do. We have all this subjective knowledge about our students built up over the course of the year. and we want to see if that subjective knowledge matches the "objective truth" of the percentage. If it matches, we feel good—our subjective evaluation has been validated. But what happens if it doesn't? What do you do when you have a student who you know has been working very hard all term—doing extra work, coming in for help, et cetera—and they come up one percentage point below some threshold for a specific letter grade? Do you go back and play with the numbers a little bit to get the "objective" grade to match your subjective knowledge? If you do, you are implicitly recognizing the myth of objectivity. Even if you do nothing, but you feel bad about the misalignment between the objective grade and your subjective knowledge, you are grappling with the inherent flaws within the point gathering paradigm.

Romagnano (2001) further argues his point by drawing on both first- and second-hand data to show how inconsistent grading is on everything from a quiz all the way up to the SAT-I mathematics test. This inconsistency creates what is called in the sciences a measurement error. For example, the measurement error on the SAT-I mathematics test is 30 points. This means that if a student scores 470, we can say with 95% certainty that their score is somewhere between 410 and 530 points. This is a huge measurement error. And it is on the SAT-I, one of the most tightly controlled assessments in the world.

What, then, does this say about measurement errors on classroom tests? Assuming that there are questions in that test wherein partial credits are given out, then there exists measurement error. To illustrate this Romagnano presents a student's solution to a factoring quadratics question and discusses how a teacher may grade it. When he asks multiple teachers to grade this same solution out of 5, their responses are equally distributed across scores 2, 3, and 4.

> This 40 percent variation is attributable to judgments that individual teachers made about the relative importance of each aspect of this student's work

described previously. In other words, these scores are subjective. (Romagnano, 2001, p. 32)

These scores are subjective. Objectivity is a myth. Yet, the myth endures. And it leads to what I have come to call the *tyranny of objectivity*, which is the harm that we do with the points we have recorded for each student across a number of events. Believing that the points we record in our gradebooks are objective, we then further believe that the sum of these points convey truth—truth about what our students have learned. But this, too, is a myth.

Ken O'Connor (2009) illustrates this best by positioning grading in the context of sky diving—my own version of which goes something like this.

> Let's pretend that we are going sky diving at your local sky diving center. After the orientation and after you have signed all the waivers, you get to pick the employee that will pack your parachute. To help you make the decision, you are provided with the parachute-packing scores for each employee across several different tests over time (see Figure 14.1). Which employee do you want to have pack your parachute?

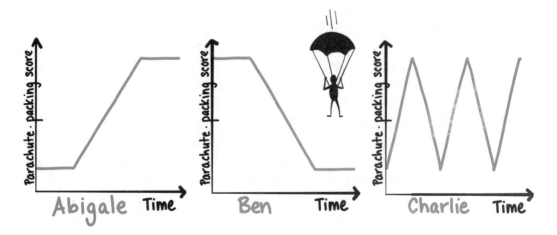

Figure 14.1 Parachute packers.

The obvious answer, of course, is Abigale. Although she was not good in the beginning, she showed steady improvement and is now consistently performing very well. Ironically, had the sky diving center treated these scores from a point-gathering perspective and provided us only with their final grade, the three employees would all have had the same grade.

The Data-Gathering Paradigm

> The fact that she didn't know how to do something in the beginning is expected—she is learn*ing*, not learn*ed*, and she shouldn't be punished for her early-not-knowing.

What O'Connor is highlighting with the parachute example is the difference between the point-gathering paradigm and what I call the *data-gathering* paradigm. Whereas the point-gathering approach would give the parachute packers all the same grade, analyzing this information as data gives us a different understanding of what has changed over time. If these three graphs were the data for three students in your class, we would say that Abigale has improved—she has learned. Isn't this what we want from our students? Isn't this our job as teachers—to help students learn? More than that, Abigale has shown mastery. Her success as a learner, and your success as a teacher, should be recognized in her grade. The fact that she didn't know how to do something in the beginning is expected—she is learn*ing*, not learn*ed*, and she shouldn't be punished for her early-not-knowing.

If we look at Ben and Charlie through a data-gathering lens as well, we quickly discern that something tragic has happened to Ben part way through the unit. His parents got divorced, or his grandmother died, or he got caught up with the wrong crowd. The same can be said about Charlie. Her inconstancy is cyclic. What is happening? I taught a Charlie once—her parents were divorced, and every two weeks she alternated who she lived with. In the weeks when she lived with one parent, she performed much better than the weeks when she lived with the other parent. The tyranny of objectivity that comes from point gathering ignores the very human elements of grading that are clearly evident in Ben and Charlie's cases. We need to let the data talk to us, rather than allowing points to rule us. The data show us where the learning is and where the problems are. Looking at this as data, rather than points, allows us to ignore early-not-knowing, to ignore outliers, and to fully acknowledge mastery when it occurs. The data-gathering paradigm is more commonly known as *outcomes-based* or *standards-based* assessment, or *evidence-based* grading, and is widely recognized as a more accurate, fair, and relevant way of grading.

TOWARD A THINKING CLASSROOM

If you want to start valuing the day-to-day evidence of learning you are witnessing in your thinking classroom and/or you wish to start evaluating students in groups, then you are going to have to make a paradigm shift. What you want to do is not feasible within a point-gathering system. You are going to have to start using a data-gathering system to get there. Whether you are just now ready to make this paradigm shift, or you made it a long time ago, the next challenge is how to capture data from all these different sources and record it in such a way that observational data can be integrated with test data and group data can be integrated with individual data.

> If you want to start valuing day-to-day learning evidence or how students work in groups, you'll have to make a paradigm shift.

In essence, the question we are trying to answer is, *How do we know where students are in their learning?* This question, it turns out, is very similar to the question we answered in the previous chapter—*How can we help students to know where they are and where they are going?* And the answer, it turns out, is the same—we need to use an instrument that delineates the learning outcomes and differentiates between the complexity levels of any given task (see Figures 14.2 and 14.3).

REPEATING PATTERNS	BASIC	INTERMEDIATE	ADVANCED
Identify the core of a repeating pattern			▓
Transfer a repeating pattern		▓	▓
Extend a repeating pattern			▓
Fill in the blanks in a repeating pattern			
Create a repeating pattern			▓

Figure 14.2 Instrument for recording student data on a repeating patterns unit.

FRACTIONS	BASIC	INTERMEDIATE	ADVANCED
Definitions			
Add and subtract proper fractions			
Add and subtract mixed fractions			
Multiply and divide proper fractions			
Multiply and divide mixed fractions			
Solve order of operation tasks with proper and mixed fractions			
Solve contextual problems involving fractions			
Estimate solutions for problems involving fractions			

Figure 14.3 Instrument for recording student data on a fractions unit.

These elegantly simple instruments allow us to record all of our data in one place. Whether those data come from an observation, conversation, or a test, they are just data points within this table. And whether a data point comes from a student doing something individually or in a group, it is just a data point in the table. We don't need to wrestle with what part of a group grade should be assigned to each student, or how much partial credit someone got on Question 5 on the test. None of that matters. We are just gathering and recording data using the same six symbols as in the previous chapter: ✓, ʃ, н, Ꮑ, ✗, and ℕ.

✓ is used when knowledge has been demonstrated individually

ʃ is used when knowledge has been demonstrated individually, but with a silly mistake

н is used when knowledge has been demonstrated individually, but with help from the teacher or a peer

Ꮑ is used when knowledge has been demonstrated within a group

✗ is used when a question has been attempted, but answered incorrectly

ℕ is used when a question has not been attempted

Add to this the subscripts **o** and **c** for whether the data comes from an observation ($\checkmark_o$) or a conversation ($\checkmark_c$), and the data can now come from the day-to-day classroom evidence of learning. This is not to say that it cannot also come from a test or a quiz but, like the practice test in Chapter 13, a test or a quiz is no longer an aggregated instrument for which a single grade is produced. Rather, a test is a collection of discrete opportunities for students to demonstrate learning, and these demonstrations are recorded on the table in a disaggregated format.

Once this organizer is populated with data (see Figures 14.4 and 14.5), you can use it to report out anecdotal comments, grades, or even percentages. For anecdotal reporting, the instrument easily and clearly helps you to structure your comments. For the student whose performance is recorded in Figure 14.4, you may say something like, "Benjamin is able to identify the core of basic repeating patterns but needs support to do so for more complex patterns," and/or "Benjamin is working toward being able to independently extend intermediate level patterns," and/or "Benjamin has repeatedly demonstrated the ability to fill in the blanks in basic and intermediate patterns, but is still working on being able to consistently do so for advanced patterns."

BENJAMIN

REPEATING PATTERNS	BASIC	INTERMEDIATE	ADVANCED
Identify the core of a repeating pattern	$\checkmark\checkmark$ S $\checkmark$	X N H G$_o$	
Transfer a repeating pattern	X $\checkmark_o$ $\checkmark$		
Extend a repeating pattern	X $\checkmark_o$ $\checkmark$ $\checkmark$	N N H H	
Fill in the blanks in a repeating pattern	$\checkmark_c$ $\checkmark_o$ $\checkmark$	X G $\checkmark_o$ $\checkmark$	$\checkmark$
Create a repeating pattern	$\checkmark\checkmark$	$\checkmark\checkmark\checkmark\checkmark$	

Figure 14.4 Benjamin's performance on the repeating patterns unit.

Creating a Grade

To turn these records into a grade or a percentage requires us to decide how a student has performed on each outcome and give that performance a numerical value. We found that, in order to determine this value, we needed to follow two foundational principles:

1. Performance at the basic level is considered minimal attainment of the outcome. This means that a student who is able to show attainment at the basic level of each outcome would be deemed to have passed the unit.

2. The different attainment levels (basic, intermediate, and advanced) are backward compatible. This means that if a student can demonstrate attainment at the advanced level, then it is assumed that they have attained the basic and intermediate level. For example, if a student can add and subtract fractions with different denominators (advanced) it is assumed that they can also add and subtract fractions with the same denominator (basic).

In this system, adhering to the first principle results in the demonstration of basic level being worth 2 points, intermediate is worth 3 points, and advanced is worth 4 points. Using these particular point values gives a student who has demonstrated basic level attainment—and only basic level attainment—2 out of a possible 4 points, which, when converted to a percentage, is 50%, which is a pass.

Adhering to the second principle means that the number of points a student receives for a particular outcome is determined by the highest complexity level they demonstrated—irrespective of what the data says about their performance at a lower complexity level. For example, if you have evidence that a student has mastered adding and subtracting fractions with different denominators, they receive 4 points even if you have evidence that they previously struggled with adding and subtracting fractions with the same denominator. As mentioned earlier, this shows that they have learned, and their learning and your teaching needs to be acknowledged—and celebrated.

The bigger question is what it means to demonstrate attainment within a certain complexity level. To answer this, we ran several experiments where we compared an individual student's assessment data (recorded in instruments such as those in Figures 14.2 and 14.3) to their teacher's subjective impression (as gleaned from day-to-day observations of and conversations) of that same student. We did this with over 40 teachers and hundreds of students. What we were looking for was correspondence between the data and the subjective impressions of teachers. In particular, we were looking for the quantity and quality of data that was necessary for there to be correspondence. What we learned was that there was a delicate balance between too little and too much data.

BENJAMIN

REPEATING PATTERNS	BASIC	INTERMEDIATE	ADVANCED
Fill in the blanks in a repeating pattern	$\checkmark_c \checkmark_0 \checkmark$	X G $\checkmark_0 \checkmark$	$\checkmark$
Create a repeating pattern	$\checkmark\checkmark$	$\checkmark\checkmark\checkmark\checkmark\checkmark$	

Figure 14.5 Benjamin's performance on the repeating patterns unit.

For example, if Benjamin demonstrated attainment at the advanced level only once (see Figure 14.5), that corresponded with the teacher's subjective assessment of that student's attainment at that level less than 60% of the time. Conversely, if performance at the intermediate level was demonstrated five times (see Figure 14.5) it corresponded with the teacher's subjective assessment 100% of the time. But it was felt by both the researcher and the teachers that this was too much evidence—there was too much redundancy. It turned out that tipping point is two consecutive demonstrations of attainment. That is, two positive data points were sufficient to match with teachers' subjective assessments of a student provided that the two positive data points were consecutive. So, whereas ✓ ✓ was enough to show attainment, ✓ ✗ ✓ was not—more data may be needed.

ALICIA

FRACTIONS	BASIC	INTERMEDIATE	ADVANCED	OUT OF	MARK
Definitions	✓✓			2	2
Add and subtract proper fractions	✓✓	✓✓	✓✓	4	4
Add and subtract mixed fractions	✓X✓	✓SX✓✓	S✓✓	4	4
Multiply and divide proper fractions	XX✓✓	NNX✓X	✓✓✓	4	4
Multiply and divide mixed fractions	XX✓✓	XS	XXH✓✓	4	4
Solve order of operation tasks with proper and mixed fractions	XS	NNX	✓✓	4	4
Solve contextual problems involving fractions		N✓✓	✓XSX	4	3
Estimate solutions for problems involving fractions	XXN✓	XN✓S	✓✓✓	4	4
	2	3	4	30	29

Figure 14.6   Alicia's performance on the fractions unit.

So, if we take all of this together and look at Alicia's performance on the fractions unit (Figure 14.6), we can begin to assign points for each outcome. On the first outcome, Alicia receives 2 points (out of a possible 2). On the next outcome she receives 4 points—she has demonstrated attainment at every level. For the next three outcomes she also receives 4s. Although she had a rough start for each of these, in the end she demonstrated mastery at the advanced level. The same is true of the sixth outcome—order of operations with proper and mixed fractions. Even though she was not able to demonstrate attainment at either the basic or intermediate levels, on the unit test she answered two of the advanced level questions for this outcome. That is to say, for this outcome, we can ignore all the early-not-knowing that Alicia demonstrated and celebrate the fact that she learned it in the end. On the seventh outcome, her performance is rather inconsistent. Even though she has shown that she can, from time to time, answer advanced-level questions, the highest level of attainment demonstrated is at the intermediate level. Therefore, she receives a 3 for the seventh outcome. If these data were collected prior to the reporting cut off, you may wish to gather some more data on Alicia for this outcome, and the evidence that is in Figure 14.6 may actually be a result of such efforts. For the last outcome, she again had a rough start, but of late has been performing well and receives 4 points. Taken together, Alicia has been awarded 29 out 30 possible points for this unit, which can then be translated into a percentage, a letter grade, or a level, depending on what your reporting mandate is.

> We can ignore all the early-not-knowing that Alicia demonstrated and celebrate the fact that she learned it in the end.

With all this assignment of points you may be wondering if we are back in the point-gathering paradigm. The answer is no. There is no gathering of points here. There is only an analysis of data, the results of which are recorded as points only at a time mandated for reporting out. Until that time, the data live as data. If you are fortunate and work in a jurisdiction where reporting out is by spreadsheet, then a cleaned up version of the table in Figure 14.6 would be sent home indicating the level of complexity Alicia has achieved for each outcome.

Incidentally, had we been recording Alicia's progress through a point-gathering paradigm, she would have collected 35 points out of a possible 67 points, which would have translated to a much lower grade on her report card. Alicia's is an extreme case, but in our research we did this type of comparison for students in many different classrooms. In essence, what we did was take test and quiz data that teachers had recorded in an event-based gradebook and transposed them into instruments similar to those in Figures 14.2 and 14.3. We then gave these anonymized grids back to that student's teacher and asked the teacher to decide what they believed that student deserved. In approximately 80% of the cases, the teacher awarded a grade that was 10%–15% higher than they had originally awarded through their event-based gradebook. The reorganization of points into data allowed the teachers to let go of outliers and early-not-knowing.

> The reorganization of points into data allowed the teachers to let go of outliers and early-not-knowing.

Even in cases such as Jennifer's (see Figure 14.7), there was a significant improvement in the grade assigned from looking at her performance as points versus data. In a point-gathering system, Jennifer would have received 28 points out of a possible 76 points. This will result in a failing grade regardless of what kind of scaling is applied. In a data-gathering system, however, Jennifer receives 22 out 30. This is a pass, and in some jurisdictions will result in a B or a C+.

JENNIFER

FRACTIONS	BASIC	INTERMEDIATE	ADVANCED	OUT OF	MARK
Definitions	√X√ₑ√₀			2	2
Add and subtract proper fractions	√√	XXH√ₑ√	XXNX₀X₀	4	3
Add and subtract mixed fractions	NXH√X	√ₑ√√	XGHX	4	3
Multiply and divide proper fractions	√√	XS√ₑ√	XNGX₀	4	3
Multiply and divide mixed fractions	XH√ₑ√√	XHHX	NNX	4	2
Solve order of operation tasks with proper and mixed fractions	√S√ₑ	XHX₀	XNN	4	2
Solve contextual problems involving fractions		XH√ₑ√₀	GXHX₀	4	3
Estimate solutions for problems involving fractions	√√	XH√	N√ₑ√√	4	4
	2	3	4	30	22

Figure 14.7   Jennifer's performance on the fractions unit.

There are two main reasons that contribute to this disparity between Jennifer's two possible grades:

> The point-gathering system repeatedly punishes Jennifer for not being able to do things she just cannot yet do.

1. The amount of data (or points) gathered at complexity levels that she just was not going to achieve. A total of 23 data points exists at the advanced level in outcomes 2–7. The point-gathering system repeatedly punishes Jennifer for not being able to do things she just cannot yet do. The data-gathering system ignores these.

2. The extent to which Jennifer's teacher was willing to look for evidence of Jennifer's learning. When she found areas that Jennifer was not performing well on, she worked with her (H), had conversations (G) with her as she tried it on her own, and tried to observe (o) her when she was solving tasks on her own or in groups. In fact, there are 15 instances where Jennifer's teacher gathered data through conversations or observations. This is time intensive. But she does not need

to do this for all her students. Most students will be like Alicia, and most of the data will come from more formal assessments like tests or quizzes. But you have students like Jennifer, students for whom you may need to spend more time. A data-gathering approach coupled with the instruments presented in this chapter allows for this.

These improvements, in concert with the those created by the changes presented in Chapter 13, resulted in a fundamental transformation of student performance in the thinking classroom. Not only do students now know where they are and where they are going, but teachers also have a clearer and cleaner picture of where students are vis-à-vis the expected outcomes of a curriculum. And when students know as much as you do about where they are and where they are going, an interesting thing begins to happen—they start thinking about their learning rather than their grades and, as they do so, grading becomes a byproduct of learning rather than the objective of learning.

> Students start thinking about their learning rather than their grades and, as they do so, grading becomes a byproduct of learning rather than the objective of learning.

FAQ

Q You mentioned that there was a 10%–15% increase in the grades of 80% of students. Why only 80% of the students?

A Some of the 20% of the students for whom this did not make much difference are already at the top of your class. This is not to say that they won't improve through this paradigm shift; they just won't improve by 10%–15%. The rest of the students for whom this shift did not make a difference defy description. It just turns out that the particular distribution of their data tells the same story in either paradigm.

Q Isn't this just a type of grade inflation?

A Not at all. By shifting to outcomes-based assessment, you avoid both the myth and the tyranny of objectivity and start to give grades that are more accurately a reflection of what your students have learned. Adherence to event-based grading, if anything, is a form of grade *deflation*.

Q I work in a setting where the basic level of attainment is deemed to be equivalent to 60% or 70%. How do I show that?

A The easiest way to do this is to change the values of the three columns from 2, 3, 4 to 3, 4, 5 or 5, 6, 7. This will position the basic level of attainment at 60% or 71%, respectively.

Q I work in a jurisdiction where we have to report out on students' demonstrated ability to solve *knowledge* questions, *application* questions, and *thinking* questions. How do I use a data-gathering system to do that?

A Knowledge, application, and thinking—like basic, intermediate, and advanced—are backward compatible. That is, for example, if a student can do a thinking question around multiplying two-digit numbers, then they must also have the knowledge to do so. Any time you have three complexity levels that are backward compatible, then you can apply the data-gathering system as it is presented in this chapter.

Q In the example you provide in Figures 14.6 and 14.7, all of the outcomes are weighted the same. But not all outcomes should be weighted the same. How does that work when you are trying to generate a grade?

A If you think some outcomes should be worth more than others, you can add a scaling factor to them (see Figure 14.8). So, for example, if you think that definitions (Outcome 1) are worth only one point, then set it that way. Likewise, if you believe outcomes 7 and 8 are worth twice as much as the other outcomes, those outcomes are now out of 8. The easiest thing to do is to still evaluate them out of 4 and then double the points. The total for the unit would then be out of 37 rather than 30 as in Figures 14.6 and 14.7.

FRACTIONS	BASIC	INTERMEDIATE	ADVANCED	OUT OF	MARK
Definitions				1	
Add and subtract proper fractions				4	
Add and subtract mixed fractions				4	
Multiply and divide proper fractions				4	
Multiply and divide mixed fractions				4	
Solve order of operation tasks with proper and mixed fractions				4	
Solve contextual problems involving fractions				8	
Estimate solutions for problems involving fractions				8	
				37	

 In the example you provide in Figures 14.6 and 14.7, all of the data are seen as equally weighted. I don't think this should be the case. I want to value test data more than observational data. How do I do that?

This your prerogative. But I caution against blindly privileging some data over others across all students. This needs to be done on a student-by-student basis. When you wish to do this, I agree that it would be important to be able to see the differences in the data. This can be accomplished in two ways. One, you could color code your data such that one color is used for test and quiz data, one for observations, one for conversations, et cetera. The second method is to split each cell into two rows and record test and quiz data on one row and all other forms of data on the other.

I feel like this way of grading will take more time. Is that the case?

Not at all. It is true, however, that it takes a bit more time to record the data than in a point-gathering gradebook, but this is more than made up for by two time savers. First, you are no longer grading with partial credits, which saves a lot of time. Second, you will grade less. When we mapped data from the event-based gradebook to a grid similar to the one in Figures 14.2–14.8 and gave it back to the teachers, every teacher made the same comment—I have *way* too much data. And it was true, there were a lot of redundant data, most of which represented unnecessary grading. In a data-gathering paradigm you collect fewer data and use only the data you need to fill in missing information for the grade.

Q Does this mean that I may not grade every item on a test?

A Correct. If, by the time a test is taken, a particular student only needs to show you evidence of being able to do the advanced types of questions, then those are the only ones you need to grade. Indeed, that student should know these are the data you are looking for and may choose to only do those questions. As mentioned, a test is no longer an aggregated event, but a convenient venue for gathering data on a variety of outcomes at a variety of complexity levels.

Q Does this mean there will not be a grade at the top of the test when it is returned to students?

A Correct. As mentioned, a test is no longer an event. Rather, it is a collection of opportunities for students to evidence their learning. The data from the test live in a disaggregated form in your instruments and the students' navigation instruments. It is only aggregated into a grade when you have analyzed the data.

> A test is no longer an event. Rather, it is a collection of opportunities for students to evidence their learning.

Q So, if a student has already evidenced attainment for every outcome at every level, they don't need to take the test?

A Correct. Or, if you have a student for whom you already know the test will not give an accurate measure, you may choose to let them avoid it and base your grade solely on observations and conversations.

Q I have students who are operating at a much lower level than others in the class, how does this method of grading help me?

A If you have a student that is still working on demonstrating their abilities on the basic level for each outcome, then give them only that page of the test. If they complete it, you can ask them if they would like the next level, and so on. This means that you need to set up your tests to allow for this. If you do, this will help all your students know what part of the test they should be focusing on.

Q I have heard that many teachers who are doing thinking classrooms do group tests. How does that work?

A How group testing looks can really vary. Some teachers make all quizzes group quizzes. Some teachers allow students to collaborate for select questions on a test or even all of a test. This collaboration can be on whiteboards or at desks and can result in a single test being produced by a group or, after collaboration, each student writing and submitting their own test paper. Any and all of these combinations are great. They substantially reduce student anxiety and value the collaborative work that happens day-to-day in a thinking classroom. And inside of a data-gathering paradigm, it is easy to record it as a group effort (G) and make sense of at reporting time.

Q Should I count data gathered in groups (G) as equivalent to data gathered individually?

A This is up to you. Most teachers would view success in a group followed by success individually (G✓) as equivalent to consecutive individual success (✓✓). There is no contradiction here. However, if you have a positive group performance followed by an unsuccessful individual performance (G✗), this is a strong indicator that more data are needed before you can say, with certainty, that individual attainment has been achieved.

Q I agree with everything that is in this chapter and I am ready to make the paradigm shift. And the jurisdiction I work in is also encouraging us to move toward outcomes-based assessment. At the same time, however, they have a mandated online event-based gradebook that we have to use. How do I navigate that?

A I have worked with many teachers who experience this exact same situation. I call this contradictory jurisdictional mandate the *two-headed monster*, and it extends to all sorts of internal contradictions that exist within educational systems. This particular contradiction exists because your jurisdiction does not recognize

Outcomes-based grading and events-based grading are actually paradigmatically different, and not just different ways to do point gathering.

that outcomes-based grading and events-based grading are actually paradigmatically different, and not just different ways to do point gathering. My first suggestion to resolve this is to find out the limits of the mandate for using the online gradebook. If it is just to report out final grades, you will be fine. If it is to keep regular records, then one of the things you can do is set each column to an outcome, weight it to zero, and record whether a student has achieved basic, intermediate, or advanced levels inside it. You can record these using 2, 3, and 4 if you wish.

Q In my jurisdiction we are mandated to triangulate our data and gather grades of how students are performing through the framework of conversations, observations, and products (COP). Although what is in this chapter helps me in this regard, it appears that some students' grades may still end up being the result of only products (tests).

A The COP framework you speak of is a huge step forward in the way teachers are beginning to think about grading and the reporting out of grades. But, in the jurisdictions in which it is mandated, it is largely misunderstood. First, and foremost, it needs to be recognized that this framework comes out of the outcomes-based grading paradigm. Most jurisdictions that mandate this don't make this clear, leaving teachers to try to figure out how to put these data into their events-based gradebooks. As a result, I have seen many cases of teachers creating columns in their gradebooks titled "Observation" and "Conversation," in effect turning observation and conversation into events whose data are to be added to and averaged with test data. This is *averaging* data, which is not the same as *triangulating* data (see Figure 14.9).

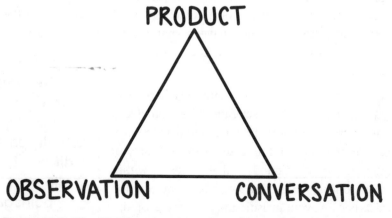

PRODUCT

OBSERVATION   CONVERSATION

Figure 14.9 The COP framework as triangulation of data.

Triangulating data means that we are gathering data from multiple sources in order to seek correspondence within the data. Correspondence indicates that you are getting close to the truth about what a student has learned. So, if a student's performance on a test corresponds with the evidence gathered in an observation, you have correspondence and you are close to having a true measure of what that student knows. Once you have correspondence between two forms of data, then you can stop. Additional data from the third data form will either be redundant and unnecessary or an outlier you can disregard. Only if the first two data forms do not produce correspondence will you need to go to the third.

> *Triangulating* data means that we are gathering data from multiple sources in order to seek correspondence within the data. Correspondence indicates that you are getting close to the truth about what a student has learned.

So, if the correspondence is between a test and a conversation, the final grade will be equivalent to that of the grade the student received on the test. This is not to say that all the data come only from the test—it only means that equivalence, as a product of correspondence, exists.

Q Does this mean that I can report out on data that come only from conversations and observations?

A Yes. Both the instruments presented in this chapter and the COP framework would allow that to happen. In fact, if you have a student who you know underperforms on tests, you can bypass the test altogether for that student.

Q I work in a jurisdiction where we are not allowed to include group work in a student's final grade. But they require us to use the COP framework. In a thinking classroom, all observational data come from a group setting. How do I negotiate this?

A First, I have a tough time imagining worthwhile observational data that do not come from a collaborative setting. Without the conversations that take place in a group, all we could do is observe a student silently doing something, the understanding behind which would be completely invisible to us. Second, your mandate likely does not prevent you from gathering group-generated data, only reporting out on it. Because you are seeking correspondence, and correspondence is, by definition, a type of redundancy, this is then an easy problem to resolve. If you have correspondence, for example, between a group observation and an individual test, then you report out only the grade the student received on the test. If the correspondence is between group observations and individual conversations, then you report out only on the conversations.

Q Group tests and quizzes sound great, and I really like the COP framework, but my students have to eventually take an external standardized test set by the regional or federal government. If I use group tests as well as gather data through observations and conversations, will my students be prepared for this?

A Students have been in several math courses before yours. And they are currently in many other courses aside from your math course. In most of those other courses, students have written individual tests. So, even if you do not do a single individual test in your course, it is highly likely that your students would still be familiar with test taking. This is not to say that you cannot help them be better prepared by giving individual tests. Nothing I have said in this chapter implies that you should not do this. I have simply provided you with an instrument that can be populated with test data (individual or group) as well as observational and conversational data. How you populate it is up to you. My only caution is that you make your choices based on what you think is best for your students and not based on the looming tyranny of an external standardized test.

Q At the end of the previous chapter, you mentioned that students start talking about what outcomes and to which levels they still need to provide evidence of competency. How do students provide this evidence?

A You have a choice—either you go get the evidence by asking a student to demonstrate something to you, or you tell the students that it is their job to bring you the evidence of what they can do. The first option is easy to do and really just involves you giving each student a customized test prior to reporting out. The second one is even easier and best achieved through the use of portfolios. You may have tried portfolios in the past and have found them clumsy, but that is because they make no sense in a point-gathering paradigm. Portfolios come out of the data-gathering paradigm, and they are just a place where students place evidence of what they understand and are able to do. Digital portfolio tools such as FreshGrade or SeeSaw are built specifically to help you and your students gather evidence and link it to specific outcomes.

Q My students are motivated by grades. You are implying that outcomes-based assessment will shift that motivation toward learning. What does it matter? Don't both result in the same thing?

A No. Outcomes-based assessment produces very different behaviors in students. Darien Allan (2017) found that students who were motivated by grades acted on that motivation only when points were on the line—their actions were discrete. On the other hand, students who were motivated to learn acted on that motivation at all times—their actions were continuous. That is, students who were motivated to learn used every opportunity to learn. And thinking classrooms are all about giving students opportunities to learn. If a student only cares about grades, they are not going to seize on these opportunities in the way that students who care about learning are.

Q In several places in this chapter you talk about *getting more data.* I understand that when the data are inconclusive, we need more data, but my school has a policy against retests. How do I then get more data?

A There are several answers to this question. First, there is no such thing as a retest in a data-gathering paradigm. The data that come from such an event does not *re*-place data, as is implied by a *re*-test. Having students write an additional test adds to the data—it doesn't replace it. So, you can say that you are not doing a retest, just an additional test. Second, additional data can come from an observation, a conversation, or asking an individual student to do a specific question for you. At some point, however, you must issue the report card. It is then, and only then, your efforts to resolve inconclusive data need to stop, and a decision needs to be made.

Q I like everything I have read, but I am struggling with a residual feeling that outcomes-based assessment is not fair. If every student's grade is based on fundamentally different data, how can we say that every student was graded the same?

A We can't. That's the whole point. We want to grade based on what the evidence says each student is capable of and, because all students are different, they will demonstrate their capabilities at different times and through different data. We accepted the idea of *differentiated instruction* a long time ago

> We accepted the idea of *differentiated instruction* a long time ago because we recognized that all students are different. If this is true, then we must also accept the idea of *differentiated assessment.*

because we recognized that all students are different. If this is true, then we must also accept the idea of *differentiated assessment*.

Your residual feelings of fairness come from a place in our collective history where assessment was a way to rank students. Some cultures still do this. And in those cultures, equality of opportunities has to be maintained in order to accurately produce the ranking. In most places in the world, however, assessment has moved from comparing students to each other to comparing students to standards (or outcomes). When that shift happened, we forgot to let go of our subconscious need to make everything the same for all students.

Q You mentioned that measurement errors occur through inconsistency in grading. But the example from Romagnano (2001) was based on inconsistencies between different graders, which could explain measurement error on big external tests. On my class tests, I am the only grader. Are there still measurement errors?

A Yes. Measurement errors also occur through our own inconsistencies. We know, for example, that test papers graded early on are often graded differently than ones that are graded near the end. For some teachers the grades improve over time; for others they get worse. We also know that if we take a break in the middle of our grading, the papers that are graded prior to the break are graded differently than the ones that are graded after the break. If that break is extended and involves sleep, a meal, and some positive or negative social events, this may exaggerate the difference. And this doesn't even take into account the variance introduced by the students and the complexity of their lives. We know, for example, that some students perform better in the morning than in the afternoon, or on Mondays better than Fridays, et cetera.

Taken together, it is safe to say that there can be a measurement error on any given student's test paper at any time. Let's conservatively say that this error is only 1%. What this means is that the grade they received on the test is an accurate reflection of what that student knows to within ±1%. This is a pretty good margin of error—much less than reality. But this is on a single test. The thing about measurement errors is that, over multiple measurements, it compounds. So, if over the year, this student will take 10 tests for you, at the end of the year her grade will be accurate to within ±10%. Depending on your particular reporting scheme, this can make the difference of several letter grades. And this is assuming an unbelievably low measurement error.

 SUMMARY

MACRO·MOVE

☐ Grade based on DATA (not points).

MICRO · MOVES

☐ Create instruments that delineate outcomes and complexity levels.

☐ Weight your outcomes.

☐ Gather observational and conversational data.

☐ Grade based on what the data are telling you.

☐ Be willing to ignore outliers and early·not·knowing.

☐ Organize your tests so that all basic questions are on the first page, etc.

☐ Let students pick which page of a test they need to do.

☐ Introduce some form of collaborative testing.

☐ Set up portfolios as a way for students to evidence their learning.

☐ Allow some students to not have to take tests.

QUESTIONS TO THINK ABOUT

1. What are some of the things in this chapter that immediately feel correct?

2. In the FAQ I distinguished between grade inflation and grade deflation. Which do you think is the bigger problem?

3. Does outcome-based grading really produce grade inflation?

4. In this chapter I gave an example of a two-headed monster that exists in some jurisdictions. My experience is that, when it comes to grading, all jurisdictions have a two-headed monster of some kind. What are the two-headed monsters you have to live with?

5. Can you think of some students for whom giving only the first page of a test could be beneficial? In what ways would it benefit them?

6. Can you think of some students for whom not writing the test could be beneficial? In what ways would it benefit them?

7. Can you think of some ways in which you could introduce collaborative testing into your assessment routines?

8. What are some of the challenges you anticipate you will experience in implementing the strategies suggested in this chapter? What are some of the ways to overcome these?

☑ TRY THIS

Start using the navigation instrument (Chapter 13) you created for your students for the next unit of study as a data-gathering instrument for yourself. Gather data throughout the unit using the six indicators (✓ ,S, H, G, X, and N) and two subscripts (o and c), and make decisions about grades based on what the data are telling you.

CHAPTER 15

PULLING THE 14 PRACTICES TOGETHER TO BUILD A THINKING CLASSROOM

• •

When I began this journey, my initial thoughts were that getting students to think is all about the tasks. If I just had the right tasks, all else would follow. Despite my experiences with Jane, all those years ago, I still believed that if we want to get students to think, then all we need to do is give them something to think about. I was both right and wrong in this thinking. Yes, if we want students to think then we need to give them something that will engage and propel them to think. But, this is far from enough. If nothing else in our practice changes, then thinking tasks will just frustrate the students and aggravate the teacher. We have to also create a culture where thinking is not only valued but also necessitated—we have to build a thinking classroom.

If you have been following along with the book, implementing new practices as you learn them, then you have already built a thinking classroom. If, however, you decided to wait until you had finished

> Once you are familiar with the 14 practices, then the question becomes, where to start? You cannot start with all of them at the same time.

> Where you start and what you do next turns out to matter.

reading about each of the 14 practices, then the question becomes, where to start? You cannot start with all of them at the same time. This would be an impossible feat even for the most talented teacher. Even if you could, your students would not be able to adapt to all the changes coming at them at once, and so this transformation would not appear as, or feel like, a success. But, you have to start somewhere. And, in this regard, where you start and what you do next turns out to matter. That is what this chapter is about—where to start and what to do next. By the end of the chapter you will have learned the ideal sequence for implementing the 14 practices to build your thinking classroom, as well as have a clear picture of what it can look like when it is all done.

THE RESEARCH

Where to start? This became my next research question. And with this question in hand, I needed a way to try and test what would work. So, I offered a variety of professional development sessions to several hundred teachers in groups of 20 to 40 at a time. I gave each group a different subset and sequence of thinking classroom practices to implement in their classrooms, and I gathered data on what sequences worked and what sequences did not. And as I did with the thinking classroom practices, I adapted what I gave subsequent groups of teachers to try.

This is not to say that I tried every possible sequence, or that the sequences were random. From the outset, I knew enough about each practice and how they interacted with each other to know that some had to come before others. For example, I knew that flow (Chapter 9) was only possible if students were working vertically (Chapter 3) and was only manageable if students were working in groups (Chapter 2). Such connections between practices helped me avoid some ineffective sequences, but I needed to test others.

BUILDING A THINKING CLASSROOM

What emerged from this experimentation was what I came to call the *Building Thinking Classrooms Framework* (see Figure 15.1)—a pseudosequential order to follow when implementing the 14 thinking classroom practices.

> What emerged from this experimentation was what I came to call the *Building Thinking Classrooms Framework.*

- Give thinking tasks
- Frequently form visibly random groups
- Use vertical non-permanent surfaces

- Defront the classroom
- Answer only keep thinking questions
- Give thinking task early, standing, and verbally
- Give check-your-understanding questions
- Mobilize knowledge

- Asynchronously use hints and extensions to maintain flow
- Consolidate from the bottom
- Have students write meaningful notes

- Evaluate what you value
- Help students see where they are and where they are going
- Grade based on data (not points)

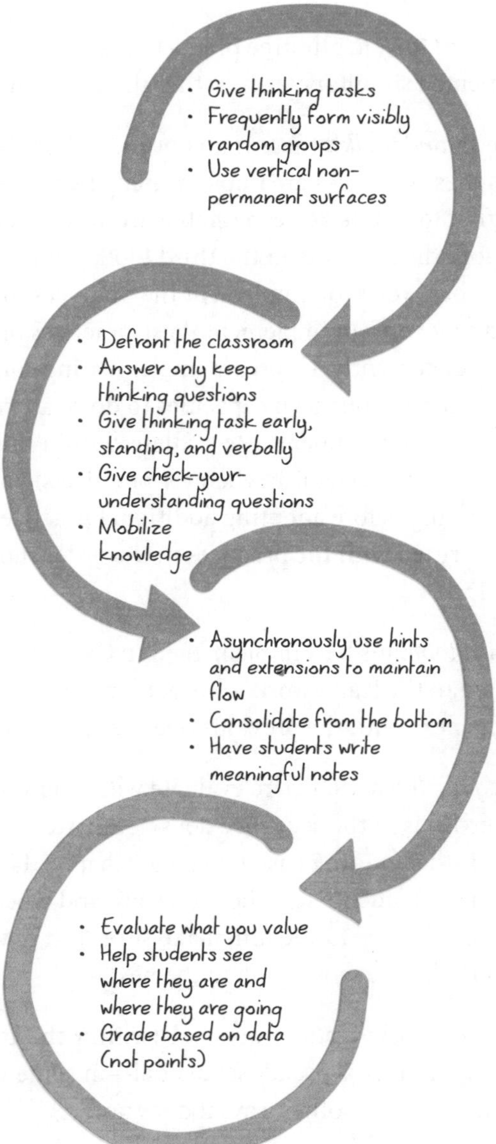

Figure 15.1 The Building Thinking Classrooms Framework.

The data showed that the 14 practices cluster into four distinct groupings, or what I call *toolkits*—and the order in which these toolkits are implemented turns out to matter. That is, the practices in the first toolkit—use thinking tasks (Chapter 1), frequently form visibly random groups (Chapter 2), and use vertical non-permanent surfaces (Chapter 3)—should have all been implemented within your classroom and working well before you move on to the second toolkit, and so on. In this way, the framework is sequential. The data also showed that within each toolkit, order also plays a role. However, not in the same way that it does between the toolkits. In particular,

1. For the *first toolkit*, all three practices need to be implemented simultaneously, rather than sequentially.

2. For the *second toolkit* there is no optimal order. That is, as long as these practices are implemented after the practices in the first toolkit have been established in your classroom and before the practices in the third toolkit, it doesn't matter what order you implement the practices in. The data showed that you can implement these practices one at a time or concurrently. If you are implementing concurrently, pay attention to your own capacities to do so as well as the capacities of your students. Regardless, within this toolkit it is important to establish whatever practice(s) you are implementing before adopting additional practices. The book is written with the practices in the order shown in Figure 15.1.

3. The *third toolkit* is best implemented in the order that it is presented in the framework—one at a time and not moving on to the next until the previous one is established.

4. For the *fourth toolkit*, where evaluate what you value (Chapter 12) is in the sequence does not matter. What matters is that grading based on data (Chapter 14) occurs after helping students see where they are and where they are going (Chapter 13). Again, make sure each practice is established before moving on to the next.

Because of these idiosyncratic sequences (within the toolkits), the framework in Figure 15.1 is pseudosequential—in some very specific ways order matters, and in other ways it does not.

Since this framework has emerged out of the research, I have been fascinated by its structure. Some parts of it make sense—like the fact that fostering autonomy should come before flow. Other things were more mysterious and have required further investigation to understand. In what follows, you'll read about each toolkit in turn and see why it is made up of the practices, and sometimes the order of the practices, that are contained within it.

Toolkit #1

To understand the practices that are embedded within the first toolkit, it helps to think about your classroom as a system, and like all systems, it operates with a rhythm of routines, expectations, and patterns that—over time—stabilize and become your classroom norms. And once these norms are established, they are very difficult to change. But you know this already. This is why you likely prefer to make changes at the beginning of the school year when the norms are yet to be established, are still in flux, and are pliable.

- Give thinking tasks
- Frequently form visibly random groups
- Use vertical non-permanent surfaces

To change the norms, then, means that you need to also change the pattern of student behaviors and habits. From system theory we know that when we try to change a stable system, the system will defend itself. "In a system, all the features reinforce each other. If one feature is changed, the system will rush to *repair the damage*" (Stigler & Hiebert, 1999). When that system is a classroom, these defenses look like resisting, complaining, and apathy.

Even worse, sometimes the changes you make are not even noticed. For example, if—part way through the year—you have ever tried to introduce journaling in the mathematics classroom, you know that it is not well received. Even if the students don't complain about it, the affordances of journaling do not initially live up to the hype. This is not to say that journaling is not good or that it is not worth doing. Journaling has been shown time and time again to be a powerful reflective tool for students as well as a way to create effective channels of communication between students and the teacher. But it usually doesn't do any of this to begin with. This is because, within a stable system, your students are unlikely to even notice that a change in their behavior is required. To them, journaling just looks like a different type of homework—which is not a radical enough departure from the norms to warrant a change on their part.

The three practices in the first toolkit, however, are not like journaling. Putting students in random groups to solve engaging thinking tasks on vertical non-permanent surfaces is enough of a departure from classroom norms that the students will notice that a change has happened. At the same time, the changes to the classroom routines are radical enough that they overwhelm the system's ability to defend itself and, as a result, the students allow themselves to change, to be different, to deviate from their normative mimicking behaviors, and to begin to really think. And so the system changes.

> The three practices in the first toolkit, when implemented together, shock the system, shocks the students, and necessitate a different behavior.

In the introduction, I told the story of how removing all the furniture had a positive impact on students' thinking. Removing furniture, it turned out, was enough of a shock to the system to change student behavior. The three practices in the first toolkit—thinking tasks, frequently formed visibly random groups, and the use of vertical non-permanent surfaces—when implemented together act in the same way. They shock the system, shock the students, and necessitate a different behavior—even if you implement it midyear.

As such, the practices in the first toolkit, despite requiring massive changes in teacher practice, are not at all about the teacher. You will either implement these thinking practices or not—that is up to you. These three practices are all about creating a new set of norms in the room that necessitate that the system change, and with it, your students.

Toolkit #2

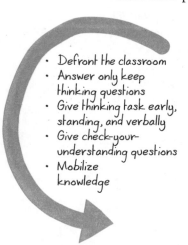

- Defront the classroom
- Answer only keep thinking questions
- Give thinking task early, standing, and verbally
- Give check-your-understanding questions
- Mobilize knowledge

Where the first toolkit is all about student behavior, the second toolkit is all about teaching practice. This toolkit will require you to make some fundamental changes to the ways in which you do things that are part of the very fabric of teaching. You will have to think about when, where, and how you give the thinking tasks, which may look very different from how you've given out tasks in the past. At the same time, you need to make conscious and deliberate changes to what kinds of questions you answer and the way in which you answer them.

> Where the first toolkit is all about student behavior, the second toolkit is all about teaching practice.

You will need to begin the process of fostering the autonomy that will allow students to manage themselves and use the groups around them as a source of hints when they are stuck and extensions when they are done. This, alone, will require you to mobilize knowledge and be deliberately less helpful—something that, for many, goes against the very foundation of what it means to be a teacher.

Also, in this toolkit is where you find the first practice that moves thinking from collective doing and knowing on vertical surfaces to individual knowing and doing through offering students a chance to complete check-your-understanding questions. Although these are seemingly like what we are used to calling homework questions, do not underestimate the changes you have to make in your own thinking about them. First, you have to give up all control over them. These are an opportunity for students and, for this reason, doing them is their responsibility—not yours.

Finally, you need to defront your classroom. Although seemingly trivial at the outset, the rearrangement of furniture will require you to use the room differently in your teaching. This rearrangement is a less efficient use of space and, therefore, you may need to get rid of unnecessary clutter, including excess furniture. It may also require you to move your desk away from the spot in the room that students most associate with the front. And it will require you to reconsider how you move around and where you stand within the classroom.

In our research, we found that if any of these practices were introduced as a first step to build a thinking classroom, they were wholly ineffective at changing student behavior. Although powerful contributors to student thinking, when introduced on their own they did not have enough impact to signal to the system that that change in student behavior was needed. Positioned here, in the second toolkit, however, these practices serve a very powerful purpose in fine tuning the thinking classroom and laying the foundation for the third toolkit.

> Positioned here, in the second toolkit, these practices serve a very powerful purpose in fine tuning the thinking classroom and laying the foundation for the third toolkit.

- Asynchronously use hints and extensions to maintain flow
- Consolidate from the bottom
- Have students write meaningful notes

Toolkit #3

Once the practices in the second toolkit have been implemented, you are ready to start creating flow in your classroom, and here is where you'll begin to truly reap the benefits of a thinking classroom. Once students are in flow, they are ready and willing to think about anything—

> Once the practices in the second toolkit have been implemented, you are ready to start creating flow in your classroom, and here is where you'll begin to truly reap the benefits of a thinking classroom.

including curriculum content. It is here that you will earn back the time you spent establishing a thinking culture in your classroom, as you start moving your students through large amounts of content very quickly. This will require more of you than anything else in the Building Thinking Classrooms Framework. Not only will you need to create sequences of curriculum tasks that allow you to walk students up the flow staircase, you will need to manage flow through the asynchronous use of hints and extensions. Your ability to give these tasks verbally will help with this, as will the groups' willingness and ability to manage themselves autonomously.

As you use this toolkit, you will also start to think deeply about how you consolidate a lesson. As mentioned in Chapter 10, planning for this begins shortly after you begin the flow activities, with the identification of student solutions that you would like to share out at the end. It also involves you seeding ideas for things that you would like to see emerge somewhere in the room. So, for example, if you would like to have a graph to share out during consolidation and no group has produced a graph, you will need to encourage one or more groups to go in this direction. All this planning and anticipation needs to happen while you are managing flow. For this reason, it is important that you have become comfortable with the asynchronous use of hints and extensions to create and maintain flow before starting to think about simultaneously planning for consolidation.

The final practice in this toolkit is meaningful notes, which serve two purposes in the thinking classroom. The first purpose is to create a record of the learning that has happened. The more important purpose, however, is to help students reify and transfer their groups' thinking into individual understanding. In this regard, meaningful notes mark, in essence, the transition of the collective consolidation that you manage into an individual record of that consolidation that the student manages. So, whether a record is important for you and your students or not, the reification ought to be. The data showed that although meaningful notes can be introduced earlier in the framework, the quality of the meaningful notes is radically improved after you have implemented consolidation from the bottom.

Toolkit #4

The final toolkit in the framework is where all the assessment occurs. This is not to say that assessment is less important than the other practices. Quite the contrary, the assessment practices are where we see some of the biggest changes in student behavior and student performance. I believe the reason these practices all ended up in the fourth toolkit is that assessment should be a reflection of a teacher's practice, and until this point, your teaching practice has been in flux. Once you have built the thinking classroom through the implementation of the first three toolkits, you are ready to start assessing in the thinking classroom.

One of the ways to do this is to start evaluating what you value. Through the use of coconstructed three-column rubrics, this assessment practice not only shows the students that you are valuing the day-to-day activities of a thinking classroom—like perseverance, risk taking, and collaboration—but also serves to fine tune student behaviors around these competencies.

Along with the use of assessment to shift student behavior vis-à-vis competencies, assessment also affords you the opportunity to shift student behavior vis-à-vis content. Through the use of the navigation instruments introduced in Chapter 13, you are able to decrypt the mysteries around content and clearly communicate to students not only the demarcations between outcomes, but also to what level they have demonstrated attainment of outcomes. The clarity of this information allows students to focus and take greater responsibility for their learning.

Once students are comfortable using these navigation instruments for data-driven self-assessment, you are ready to seamlessly use similar structures to collect and analyze data to track student performance. Making the paradigm shift to outcomes-based assessment will allow you to gather and record data not only from test sources, but

- Evaluate what you value
- Help students see where they are and where they are going
- Grade based on data (not points)

> The reason these practices all ended up in the fourth toolkit is that assessment should be a reflection of a teacher's practice, and until this point, your teaching practice has been in flux.

> Through the triangulation of data, you will be able to construct a more accurate picture of where each student is in their learning (and assign a more accurate grade) while at the same time helping students to shift their focus from grades as a product to evidence of learning as a process.

also from observational and conversational sources—both individual and collaborative. Through the triangulation of data, you will be able to construct a more accurate picture of where each student is in their learning (and assign a more accurate grade) while at the same time helping students to shift their focus from grades as a product to evidence of learning as a process.

TRANSFERRING COLLECTIVE SYNERGY INTO INDIVIDUAL KNOWING AND DOING

Early on in the research we came to a troubling realization. This was back when we were mostly exploring students working on thinking tasks in random groups on vertical surfaces and fine tuning that work with how we were giving the tasks, answering questions, and using hints and extensions to maintain flow. The research was far from finished at this point, but things were still going well. Our use of flow was still allowing us to move through huge amounts of content in very short periods of time. The problem, we realized, was that we were not seeing any improvements in students' individual performances on tests.

We knew there was likely a disconnect somewhere, and we had to figure out where it was happening. When we looked closely at groups working on the whiteboards, what we saw were students working collaboratively and clearly demonstrating understanding of the concept at hand—both collectively and individually. That is, we saw and heard individual members of groups clearly articulating and explaining their thinking in the moment. For example, when we looked closely at Grade 10 students factoring quadratics according to the flow sequence in Chapter 9, what we saw was that, in that moment, every member of a group was able to factor $8x^2 - 8x - 6$ and clearly explain what they were doing. Yet, when we tested these same students four days later, about 70% of the students were not able to factor a similar quadratic. Somehow, the knowing and doing that we were seeing in the collaborative groups was dissipating and not transferring into individual knowing and doing.

The learning we were seeing in the groups was real. But it was synergistic, temporal, and contextual—existing only in that moment, in that context, and within that collective. The individual understanding that was being demonstrated in those moments was not individual at all—it was only the individual expressions of the knowing and doing that was being held collectively within the synergy of the group in that moment. What we needed to figure out was how to transform and transfer that synergistic collective knowing and doing into the individual knowing and doing. At the time we had no idea how to do this. But amazingly, over time, and with the emergence of additional optimal thinking practices, this issue began to resolve itself. And with the emergence of the rest of the Building Thinking Classrooms Framework and its pseudosequential implementation schedule, the final pieces fell into place.

> What we needed to figure out was how to transform and transfer that synergistic collective knowing and doing into the individual knowing and doing.

It turns out that there are four thinking practices that work together to achieve this transfer from collective to individual understanding. And, at the time that we were beginning to see the transference issue, none of these were being experimented with yet. These are

- consolidation from the bottom (Chapter 10),

- meaningful notes (Chapter 11),

- check-your-understanding questions (Chapter 7), and

- helping students see where they are and where they are going (Chapter 13).

On their own and in concert, each of these contributes to transforming synergistic collective understanding into individual understanding (see Figure 15.2).

Consolidation from the bottom helps to name and formalize the synergistic experiences of the collaborative work. But this is still a form of collective consolidation. Meaningful notes provide students with the first individual opportunity to consolidate the collective learning—from the group and the teacher's consolidation—and extract from it their personal learning. Check-your-understanding questions offer them an immediate opportunity to confirm this learning. Finally, helping students to see where they are and where they are going offers ongoing and delayed formative feedback that positions this learning within the scope of the unit of study.

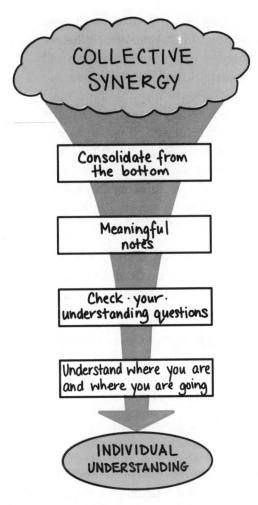

Figure 15.2 Transforming collective synergy into individual understanding.

Although each of these practices helps, in some way, to move collective learning toward individual learning, they are most effective when all four work together. As with many things in the Building Thinking Classrooms Framework, in this regard the order matters—but not in the same way order has mattered up until now. In this case the order that matters is the order in which the first three of these practices—consolidation from the bottom, meaningful notes, check-your-understanding questions—occur within a lesson (see Figure 15.3).

- You begin the lesson by giving a task verbally with the students standing around you somewhere in the room, randomly grouping the students, and sending them off to their VNPSs.

- You then manage the flow in the room by using hints and extensions, while at the same time planning for consolidation.

- You keep the students in flow until the energy wains, at which point you consolidate from the bottom.

- This is followed by meaningful notes.

- Finally, you provide the opportunity to do check-your-understanding questions.

Although each of these practices helps, in some way, to move collective learning toward individual learning, they are most effective when all four work together.

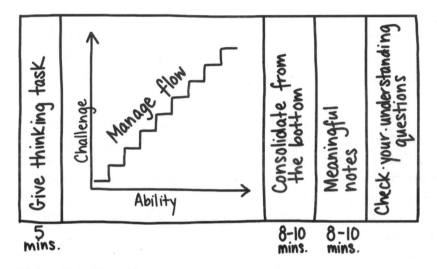

Figure 15.3 Typical lesson sequence.

This sequence and pacing works well if the lessons are at least 65 minutes long. If they are shorter than this, you may struggle to fit everything into one lesson. If this is the case, you may wish to spread these activities out over two lessons (see Figure 15.4). In this situation, Day 1 would be dedicated to having the students work in flow and may or may not end with consolidation. Day 2 would then begin with a brief statement of the task and reemergence within flow—except for a shorter period of time and moving through the sequence of tasks in bigger jumps. This is followed by consolidation—even if this was done on Day 1, meaningful notes, and check-your-understanding questions. This sequence is flexible and can take on different configurations. The only restrictions, we found, were that immersion in a thinking activity must precede consolidation, and consolidation must happen before meaningful notes. For example, Day 1 can include consolidation and meaningful notes, leaving Day 2 to be dedicated to check-your-understanding questions. In fact, many teachers I have worked with will occasionally dedicate an entire lesson to students doing check-your-understanding questions coupled with documenting where they are and where they are going.

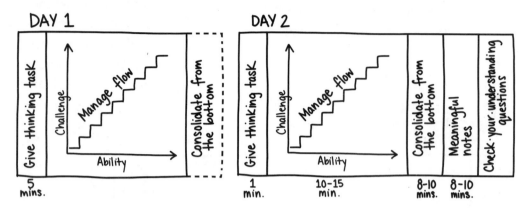

Figure 15.4 Typical lesson sequence spread over two days.

For the transfer to individual understanding to occur, students need to take on more and more responsibility for their own learning.

There is one additional thing to note about these four practices. They are not just about transferring collective synergy to individual understanding. They are also about transferring responsibility of teaching into responsibility for learning That is, for the transfer to individual understanding to occur, students need to take on more and more responsibility for their own learning. Aside from consolidating from the bottom, the remaining three practices (check-your-understanding questions, meaningful notes, and understanding where they are and where they are going) require student responsibility in abundance. This is why I call these three practices the *student-responsibility practices*—and there is a hierarchy among them.

Check-your-understanding questions require the least amount of responsibility. Students are familiar with homework. Check-your-understanding questions are just about shifting who they are done for (from teacher to students) and why they are done (from practice to checking for understanding). In essence, the only difference between homework and check-your-understanding questions is a shift of responsibility. This is why this practice appears first of the three responsibility practices, and it is the only one of the three to be included within the second toolkit.

Meaningful notes comes next—in the third toolkit. This requires more responsibility on the part of the students and is a bigger departure from what they are familiar with. Students are used to being told exactly what to write down in their notes. In normative settings, note taking is more akin to scribing what is written on the boards. This requires little to no thinking and little to no ownership over what gets written down. Meaningful notes is a massive departure from this passive activity. Not only do students now have to extract their own meaning from their collaborative activity and the teacher's consolidation, they have to choose how to sequence and represent this meaning, and they have to take ownership over *its production* in the moment and *its use* in the future.

Understanding where they are and where they are going comes last—in the fourth toolkit. This practice requires the most responsibility in that it is a call for students to completely own their learning. Although the teacher creates the structures to do this, the students ultimately have to monitor and track their learning *and* take action when they see that they are not attaining an outcome at various complexity levels.

Each of these student-responsibility practices is offered as an opportunity to students in a thinking classroom. And, as opportunities, they are sensitive to teacher meddling. That is, if the teacher makes motions to require students to take on this responsibility through grading or punitive measures, then they transform from responsibility practices to accountability practices, and we are right back to students doing them for the wrong reason (grades) and for the wrong person (teacher). I believe it is because of this sensitivity and the hierarchical increase in responsibility within these three practices that they have been distributed within the Building Thinking Classrooms Framework the way they have—one each, in order of required responsibility, in the second, third, and fourth toolkits. And I also believe that this is why there are no student-responsibility practices within the first toolkit—which is about establishing a collaborative thinking culture more than about facilitating individual knowing and doing.

REBUILDING A THINKING CLASSROOM

When you have all four toolkits and 14 practices up and running in concert with each other, and students have begun taking ownership more of their learning, you will realize the full benefits of a thinking classroom—greater student engagement, more active learning, greater student enjoyment, increased student responsibility, higher performances, easier and faster movement through content, greater satisfaction as a teacher, et cetera. And just when you are most enjoying these benefits, they change the students on you, and you have to start all over again. Or do you?

> **What is the sequence of implementation of the thinking classroom practices after Year 1?**

This became my next research question—what is the sequence of implementation of the thinking classroom practices after Year 1? Although your new class of students may not be familiar with the thinking classroom, you are. After having built and run a thinking classroom, you are well versed in each of the 14 thinking classroom practices. This means that the implementation framework would be dependent only on student acclimatization. Does that change the sequence of the practices? It turns out that it does. The data showed that for teachers going into their second, third, or fourth year of implementation, the pseudosequence looked different and was composed of only two toolkits. I called this pseudosequence the *Rebuilding Thinking Classrooms Framework* (see Figure 15.5).

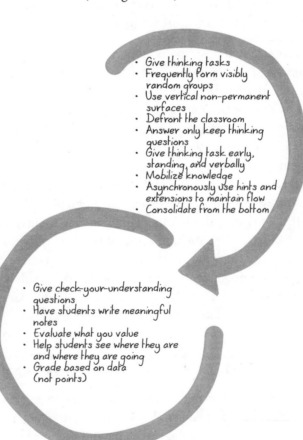

Figure 15.5 The Rebuilding Thinking Classrooms Framework.

Toolkit #1

There are several major differences between the first toolkit in this rebuilding framework and the first toolkit in the original framework for building a thinking classroom for the first time. The main difference is the number of practices. I believe that this is solely due to the fact that whereas the original framework is governed by the acclimatization periods for both students and the teacher, the rebuilding framework is governed only by the acclimatization rate of students. Because you, as the teacher, are already used to the thinking classroom practices, you are ready to go on Day 1. You have a collection of highly engaging non-curricular thinking tasks to start with, you know when, where, and how to give these tasks, and how to avoid answering stop-thinking and proximity questions. Your room is already defronted and set up with vertical non-permanent surfaces, you know how to randomize groups, and you are ready to start mobilizing knowledge by being deliberately less helpful. At the same time, you know how to sequence activities to create flow and how to consolidate the learning after a flow experience.

- Give thinking tasks
- Frequently form visibly random groups
- Use vertical non-permanent surfaces
- Defront the classroom
- Answer only keep thinking questions
- Give thinking task early, standing, and verbally
- Mobilize knowledge
- Asynchronously use hints and extensions to maintain flow
- Consolidate from the bottom

Like the practices in the first toolkit in the original framework, the practices in the first toolkit in the rebuilding framework are all implemented at once. This is not to say that this is all going to go well on Day 1. Students will be shocked at how different your class experience is from other learning experiences, and this may result in some resistance and will definitely require an acclimatization period. But, because this is being initiated on Day 1 of the school year, resistance will be minor, and acclimatization will be quick. Unlike the first time you did this, you are now in the enviable position of building classroom norms as opposed to trying to change classroom norms—a much easier task. Regardless, many of the students' thinking classroom behaviors will take to time to develop, and until they do your efforts to build and maintain flow may feel clumsy—but still worth doing.

> You are now in the enviable position of building classroom norms as opposed to trying to change classroom norms.

Toolkit #2

- Give check-your-understanding questions
- Have students write meaningful notes
- Evaluate what you value
- Help students see where they are and where they are going
- Grade based on data (not points)

After about three weeks, your class will be ready to move on to the second toolkit, which contains all of the assessment practices and all of the student-responsibility practices. Unlike what you did with the first toolkit, you will implement these one at a time as the students become ready for them. Although there is some degree of freedom around the order with which you do this, it still works best if you present the move to outcomes-based assessment to students after they have been exposed to assessment as a tool for helping them see where they are and where they are going. This is not to say that you should not be doing outcomes-based assessment from the beginning, but rather that your students will not understand how it works until they have had a chance to first experience it in the context of data-driven self-assessment.

Likewise, I found that the three student-responsibility practices need to roll out in the same order as in the Building Thinking Classrooms Framework—and for the same reason. There are differences in the amount of student responsibility each practice requires, and it is important to increase the responsibility gradually, beginning with check-your-understanding questions, and so on.

FOREST FOR THE TREES

My goal from the outset was to get students to think.

Whether you have implemented the 14 thinking practices as you read the book or you are ready to start implementing according to the Building Thinking Classrooms Framework (see Figure 15.1), it is important to try to see the forest for the trees. My goal from the outset was to get students to think (forest). Thinking is a necessary precursor to learning, and if students are not thinking they are not learning. The ensuing 15 years of research, involving hundreds of teachers and thousands of students, was singly—obsessively—focused on this goal. What emerged were 14 thinking practices (trees). This book lays out these 14 practices along with the many micropractices (more trees) that will help you in your quest to build a thinking classroom.

Keeping both perspectives in balance is necessary to your success, and thus, that of your students. It is so easy to become consumed by the importance of each practice that we lose sight of the overall objective. If we want students to think we need to give them something to think about and someone to think with and somewhere to think. In our pursuit of that goal, it is also important to not rob students of the opportunity to think by answering all their questions or telling them how to do something. Along the way we are going to have to fine tune the way our room looks, what the acts of teaching look like, and what and how we assess. And students are going to have to take more responsibility for their learning. But these are all trees. The goal is still the forest—getting our students to think.

And when you achieve this, all of these practices will lose their discreteness and meld into a whole—and you will have a classroom that is not only conducive to thinking but also requires thinking, a space that is inhabited by thinking individuals as well as individuals thinking collectively, learning together, and constructing knowledge and understanding through activity and discussion (Liljedahl, 2016). You will have built a thinking classroom—you will have your forest.

> It is so easy to become consumed by the importance of each practice that we lose sight of the overall objective.

> And when you achieve this, all of these practices will lose their discreteness and meld into a whole—and you will have built a thinking classroom.

FAQ

Q A thinking classroom lesson seems to revolve around the idea of creating and maintaining flow through the way we use hints and extensions. Given how important this is, why not place it earlier in the Building Thinking Classrooms framework?

A I agree that creating and maintaining flow is of utmost importance, and likely the most significant practice, in the thinking classroom. But I did not choose to place it in the third toolkit. The pseudosequence presented in Figure 15.1 emerged out of the research as the order most effective at building a thinking classroom. It has taken a long time for me to understand why this is the pseudosequence as opposed to others. What I present in this chapter are the results of much theorizing as to why some practices need to be implemented together and why some practices need to be established before others are adopted. In this regard, flow is where it is because a lot of the thinking practices in the first and second toolkits need to be working well within the classroom before flow will function as the research shows it can.

Q I was thinking of starting more gradually than what the sequence suggests—maybe with changing the way I answer questions or give tasks. Can I do that?

A You can do what you wish. This book offers you a set of practices that have been proven to initiate and maintain thinking. The Building Thinking Classrooms (Figure 15.1) and Rebuilding Thinking Classrooms (Figure 15.9) frameworks offer proven pseudosequences to enact these practices. You have to decide for yourself what is right for you and your students. But I caution you against starting gradually. One of the things the research showed over and over again is that if the change you make is too subtle, then the students' behaviors don't change, and the changes in your teaching practice will have very little impact. As mentioned, this is why the first toolkit in the Building Thinking Classrooms Framework contains the practices it does and why they are to be enacted together—to signal to students that this is different, and that they need to behave differently.

Q I was thinking of starting off by doing thinking classrooms once a week. Will that work?

A Doing it once a week will serve to give your students a break from the normal routines of school, but our research shows that you are unlikely to get beyond the first toolkit. It will also result in the students seeing *math class* and *thinking classrooms* as two distinct events rather than seeing math class as being about thinking in math.

Q Are there any other configurations for breaking the lesson sequence across two days other than what is presented in Figure 15.8?

A Yes there are. I gave one other configuration already—that of having Day 2 dedicated to check-your-understanding questions. Another configuration I have seen is to have students do meaningful notes after they have had an opportunity to do check-your-understanding questions. The research showed that meaningful notes need to come after consolidation—but they do not have to come immediately after. Some teachers I have worked with ask students to do their meaningful notes at home and preserve more class time for doing check-your-understanding questions. This increases the responsibility required of students around meaningful notes, but it works.

Q I found it odd that check-your-understanding questions emerge in the Building Thinking Classrooms Framework (Figure 15.1) before consolidation and meaningful notes, yet occur after these practices in the lesson structure (Figures 15. 7 and 15.8). Why is that?

A The pseudosequence of the Building Thinking Classrooms Framework is what it is for a number of reasons—from what the teacher is capable of and ready to implement to what the students are capable of and ready to take on. Check-your-understanding questions emerged in the second toolkit because of where they sit within the hierarchy of the student-responsibility practices. They sit where they do within the lesson sequence because of the empirically emergent results that meaningful notes must be preceded by consolidation.

QUESTIONS TO THINK ABOUT

1. What are some of the things in this chapter that immediately feel correct?

2. What are some different ways to split a thinking classroom lesson across two lessons?

3. Where are you in your journey through the thinking classroom framework, and what are you going to introduce next?

4. In this chapter I spoke about the classroom as a system and how systems defend themselves against change. Can you think of a time where you tried to introduce something that the system defended itself against?

5. If the best time to introduce something new to a system is in the first week of school, what do you want the start of your next school year to look like?

6. What are some of the challenges you anticipate you will experience in implementing the strategies suggested in this chapter? What are some of the ways to overcome these?

REFERENCES

Allan, D. (2017). *Student actions as a window into goals and motives in the secondary mathematics classroom* [Unpublished doctoral thesis]. Simon Fraser University.

Andrews, D., & Bandemer, K. (2018). Refining planning: Questioning with purpose. *Teaching Children Mathematics, 25*(3), 166–173.

Bettelheim, B. (1976). *The uses of enchantment: Meaning and importance of fairy tales.* Knopf.

Boaler, J. (2002). *Experiencing school mathematics: Traditional and reform approaches to teaching and their impact on student learning* (Rev. and exp. ed.) Lawrence Erlbaum Associates.

British Columbia Ministry of Education. (2020). *BC performance standards— numeracy (grade 6).* Retrieved February 23, 2020, from https://www2.gov.bc.ca/assets/gov/education/administration/kindergarten-to-grade-12/performance-standards/numeracy/numerg6.pdf

Carroll, L. (1880, February). The cats and rats again. *The Monthly Packet of Evening Readings for Members of the English Church*, Vol. 29, pp. 107–108. Walter Smith.

Clarke, D., & Xu, L. (2008). Distinguishing between mathematics classrooms in Australia, China, Japan, Korea and the USA through the lens of the distribution of responsibility for knowledge generation: public oral interactivity and mathematical orality. *ZDM, 40*(6), 963–972.

Cobb, P., Wood, T., & Yackel, E. (1991). Analogies from the philosophy and sociology of science for understanding classroom life. *Science Education, 75*(1), 23–44.

Cobb, P., Wood, T., Yackel, E., & McNeal, B. (1992). Characteristics of classroom mathematics traditions: An interactional analysis. *American Educational Research Journal, 29*(3), 573–604.

Csíkszentmihályi, M. (1990). *Flow: The psychology of optimal experience.* Harper and Row.

Csíkszentmihályi, M. (1996). *Creativity: Flow and the psychology of discovery and invention.* Harper Perennial.

Csíkszentmihályi, M. (1998). *Finding flow: The psychology of engagement with everyday life.* Basic Books.

Davies, A., & Herbst, S. (2013). Co-constructing success criteria. *Education Canada, 53*(3), pp. 16–19.

Davis, B., & Simmt, E. (2003). Understanding learning systems: Mathematics education and complexity science. *Journal for Research in Mathematics Education, 34*(2), 137–167.

Dweck, C. (2016). *Mindset: The new psychology of success.* Ballantine.

Dweck, C., & Leggett, E. (1988). A social-cognitive approach to motivation and personality. *Psychological Review, 95*, 256–273.

Edwards, J., & Jones, K. (2003). Co-learning in the collaborative mathematics classroom. In A. Peter-Koop, V. Santos-Wagner, C. Breen, & A. Begg (Eds.), *Collaboration in teacher education: Mathematics teacher education* (Vol 1., pp. 135–151). Springer.

Egan, K. (1988). *Teaching as story telling*. University of Chicago Press.

Elrod, M. J., & Strayer, J. F. (2015). Using an observational rubric to facilitate change in undergraduate classroom norms. In C. Suurtamm & A. Roth McDuffie (Eds.), *Annual perspectives in mathematics education: Assessment to enhance teaching and learning* (pp. 87–96). National Council of Teachers of Mathematics.

Esmonde, I. (2009). Mathematics learning in groups: Analyzing equity in two cooperative activity structures. *Journal of the Learning Sciences, 18*(2), 247–284.

Fenstermacher, G. (1986). Philosophy of research on teaching: Three aspects. In M. C. Whittrock (Ed.), *Handbook of research on teaching* (3rd ed., pp. 37–49). Macmillan.

Fenstermacher, G. (April 4–8, 1994, Rev. 1997). *On the distinction between being a student and being a learner* [Paper presentation.] Annual meeting of the American Educational Research Association, New Orleans, LA.

Fosnot, C., & Dolk, M. (2002). *Young mathematicians at work: Constructing fractions, decimals, and percents*. Heinemann.

Frey, N., Hattie, J., & Fisher, D. (2018). *Developing assessment-capable visible learners, grades K–12: Maximizing skill, will, and thrill*. Corwin.

Hatano, G. (1988, Fall). Social and motivational bases for mathematical understanding. *New Directions for Child Development, 41*, 55–70.

Hattie, J. (2009). *Visible learning: A synthesis of over 800 meta-analyses relating to achievement*. Routledge.

Horowitz, M. (1967). Role theory: One model for investigating the student-teaching process. *McGill Journal of Education, 2*(1), 38–44.

Jansen, A. (2006). Seventh graders' motivations for participating in two discussion-oriented mathematics classrooms. *Elementary School Journal, 106*(5), 409–428.

Kerkhoff, M. (2018). *Experiencing mathematics through problem solving tasks* [Unpublished master's thesis]. Simon Fraser University.

Kotsopoulos, D. (2007). Investigating peer as "expert other" during small group collaborations in mathematics. In *Proceedings of the 29th annual meeting of the North American Chapter of the International Group for the Psychology of Mathematics Education* (pp. 685–686). University of Nevada, Reno.

Liljedahl, P. (2016). Building thinking classrooms: Conditions for problem solving. In P. Felmer, J. Kilpatrick, & E. Pekhonen (Eds.), *Posing and solving mathematical problems: Advances and new perspectives* (pp. 361–386). Springer.

Liljedahl, P. (2018). On the edges of flow: Student problem solving behavior. In S. Carreira, N. Amado, & K. Jones (Eds.), *Broadening the scope of research on mathematical problem solving: A focus on technology, creativity and affect* (pp. 505–524). Springer.

Liljedahl, P., & Allan, D. (2013a). Studenting: The case of homework. In M. V. Martinez & A. C. Superfine (Eds.), *Proceedings of the 35th Conference for Psychology of Mathematics Education—North American Chapter* (pp. 489–492). University of Illinois at Chicago.

Liljedahl, P., & Allan, D. (2013b). Studenting: The case of "now you try one." In A. M. Lindmeier & A. Heinze (Eds.), *Proceedings of the 37th conference of the International Group for the Psychology of Mathematics Education* (Vol. 3, pp. 257–264). PME.

Lithner, J. (2008). A research framework for creative and imitative reasoning. *Educational Studies in Mathematics, 67*(3), 255–276.

Liu, M., & Liljedahl, P. (2012). "Not normal" classroom norms. In T. Y. Tso (Ed.), *Proceedings of the 36th conference of the International Group for the Psychology of Mathematics Education* (Vol. 4, p. 300).

Marton, F., & Tsui, A. B. M. (2004). *Classroom discourse and the space of learning.* Lawrence Erlbaum Associates.

Mason, J., & Pimm, D. (1984). Generic examples: Seeing the general in the particular. *Educational Studies in Mathematics, 15*(3), 277–289.

Mehrabian, A. (2009). *Nonverbal communication.* Transaction Publishers.

Merriam-Webster. (n.d.). *Merriam-Webster.com dictionary.* Retrieved March 23, 2020, from https://www.merriam-webster.com/dictionary/objective

National Council of Teachers of Mathematics (NCTM). (2000). *Principles and standards for school mathematics.* Author.

National Council of Teachers of Mathematics (NCTM). (2012). The unusual baker. *Teaching Children Mathematics, 18*(5), 278–280.

National Council of Teachers of Mathematics. (NCTM). (2014). *Principles to actions: Ensuring mathematical success for all.* Author.

O'Connor, K. (2009). *How to grade for learning.* Corwin.

Peper, E., & Lin, I. (2012). Increase or decrease depression: How body postures influence your energy level. *Biofeedback, 40*(3), 125–130.

Pólya, G. (1945). *How to solve It.* Princeton University Press.

Race, P. (2010). *Making learning happen: A guide for post-compulsory education.* Sage.

Romagnano, L. (2001). The myth of objectivity in mathematics assessment. *Mathematics Teacher, 94*(1), 31–37.

Schoenfeld, A. (1985). *Mathematical problem solving.* Academic Press.

Slavin, R. E. (1996). Research on cooperative learning and achievement: What we know, what we need to know. *Contemporary Educational Psychology, 21*(1), 43–69.

Smith, M., & Stein, M. K. (2011). *5 Practices for orchestrating productive mathematics discussions.* National Council of Teachers of Mathematics.

Smith, M., & Stein, M. K. (2018). *5 practices for orchestrating productive mathematics discussions* (2nd ed.). National Council of Teachers of Mathematics.

Staples, M. (2007). Supporting whole-class collaborative inquiry in a secondary mathematics classroom. *Cognition and Instruction, 25*(2–3), 161–217.

Stiggins, R., Arter, J., Chappuis, J., & Chappuis, S. (2006). *Classroom assessment for student learning: Doing it right—using it well.* Prentice Hall.

Stigler, J., & Hiebert, J. (1999). *The teaching gap: Best ideas from the world's teachers for improving education in the classroom.* The Free Press.

Urdan, T., & Maehr, M. (1995). Beyond a two-goal theory of motivation and achievement: A case for social goals. *Review of Educational Research, 65*(3), 213–243.

Vogler, K. (2008, Summer). Asking good questions. *Educational Leadership, 65* (online only). http://www.ascd.org/publications/educational-leadership/summer08/vol65/num09/toc.aspx

Wells, K. (2014). *A conversation–gesture approach to recognising mathematical understanding in group problem solving (teaching from the sidelines)* [Unpublished doctoral thesis]. Simon Fraser University.

Wilson, V., & Peper, E. (2004). The effects of upright and slumped postures on the recall of positive and negative thoughts. *Applied Psychophysiology and Biofeedback, 29*(3), 189–195.

Yackel, E., & Cobb, P. (1996). Sociomathematical norms, argumentation, and autonomy in mathematics. *Journal for Research in Mathematics Education, 27*(4), 458–477.

Zager, T. (2017). *Becoming the math teacher you wish you'd had: Ideas and strategies from vibrant classrooms.* Stenhouse Publishers.

Zazkis, R., & Liljedahl, P. (2008). *Teaching mathematics as storytelling.* Sense Publishers.

INDEX

marking, 127
navigation instrument for, 236,
    239, 248. *See also* Formative
    assessment
for primary grades, 129
Clarification questions, 88
Classroom
    cell phone usage in, 104
    defronted, 75–77, 75 (figure),
        77 (figure)
    furniture arrangement in. *See*
        Furniture arrangement
    journaling in, 283
    norms, 11–12
    orderly, 75 (figure)
    projector placement in, 77–78
    student workspace in. *See* Workspace,
        student
    super organized, 72
    as system, 283–284
    See also Thinking classroom
Coconstruction of rubric, 219,
    223–224, 287
Collaboration rubric, 213 (figure),
    216 (figure)
    four-column, 214–218
Collaborative groups, 39, 45 (figure)
    ability groupings, 53, 65–66
    active and passive interaction among,
        135–137, 137 (figure)
    check-your-understanding questions
        for, 128
    diversity, 40, 45
    for educational goals, 39–40
    in elementary classroom, 39
    erasing freedom in, 66
    evaluation, 223
    FAQ, 49–54
    flow in, 162–163
    formation by playing cards activity,
        44, 50–51
    frequent visibly random groupings,
        44, 46–49
    in high school, 40
    idea, borrowing, 48–49
    integration, 40
    issue, 39–40
    knowledge mobility and, 47–49
    macro-moves, 54
    marker movement in, 64–65
    mathematics learning enthusiasm, 48
    micro-moves, 54
    optimal group size, 44–45
    for peacefulness, 40

pedagogy, 39
problem, 40–42
for productivity, 39
quizzes/tests, 254, 271, 274
random grouping, 43, 50–52
redundancy, 44–45
role assignment in, 42
self-correction in, 141
self-selection, 40–42, 45
size, 44–45
social barriers, elimination of, 46–47
for social goals, 40
socialization, 40
social stress reduction, 48–49
strategic grouping, 39, 41–43
teacher goals *vs.* student goals, 40
thinking classroom implementation,
    43–49
Try This activity, 55, 68–69
turn taking, 45
visibly random grouping, 44, 46–49
willingness, 46, 50
with VNPSs, 64
with wall-mounted whiteboards,
    58–59, 59 (figure)
Collective consolidation, 289
Collective synergy to individual
    understanding, 288–293,
    290 (figure)
Competencies
    evaluation of, 211–222
    meaningful worked examples,
        199–200
    observable, 220, 224
    students' success in thinking
        classroom, 209–210
Consolidation, 171
    cell phone usage during, 176
    with change model, 173
    collective, 290
    in collective synergy to individual
        understanding transformation,
        289–293
    discussing and demonstrating, 180
    discussion with notation, 175, 182
    discussion without notation, 174, 182
    erasing option in, 179–181
    FAQ, 179–182
    gallery walk, 177–178, 180–182
    guided gallery walk, 179, 181–182
    issue, 171
    leveling to bottom approach,
        172–176, 182, 202, 285–286

complexity level, attainment within, 262–264
COP framework, 272–274, 272 (figure)
creation, 262–267
data-gathering paradigm, 258, 268, 272, 274
deflation, 268
event-based, 254–257, 265, 271–272
evidence-based, 258
failing, 265
FAQ, 267–276
formative assessment, 231–232
group quizzes/tests, 254, 271, 274
group *vs.* individual, 271
inconsistency, 256, 276
inflation, 268
issue, 253–254
macro-moves, 277
measurement error, 256–257, 276
micro-moves, 277
myth of objectivity, 255–256
objective, 255–256
online-based, 271–272
outcomes and complexity levels, instrument for, 259–261
paradigm shift, 259, 267
pass, 265
performance levels, 261
point-gathering paradigm, 254–257, 265–266, 271–272
problem, 254–258
retest, 275
standardized test, 274
thinking classroom implementation, 259–267, 287–288
triangulation of data, 272–273, 287–288
Try This activity, 278
tyranny of objectivity, 257
Graphic organizers, 195
with demarcated cells, 196, 196 (figure)
with prelabelled cells, 197 (figure), 198, 199 (figure)
with restricted cells, 195, 196 (figure)
Grouping. *See* Collaborative groups
Group quizzes/tests, 254, 271, 274
Growth mindset, 217
Guided gallery walk, 179, 181–182

Heterogeneous groups, 39
High-ceiling tasks, 23

Highly engaging thinking tasks, 21–22, 31–32
High school
cell phone usage in, 104
challenge and skill balance, 151–152
grouping in, 40
highly engaging thinking tasks in, 21–22
Hints, 145
for challenge and skill balance, 156–158
for decreasing challenge, 156–157
FAQ, 160–166
for flow, 156–158
for increasing ability, 156–157
issue, 145
macro-moves, 166
micro-moves, 166
for notes, 202–203
problem, 145–146
thinking classroom implementation, 146–160, 285–286
Try This activity, 168–169 *See also* Optimal experience
Homework, 119
assignments, 126
cheating behavior, 121–122
checking, 121
check-your-understanding questions, 125–127
FAQ, 127–129
forgetfulness, 120
getting help in, 122
issue, 119
macro-moves, 130
marking, 120–121, 123–124, 124 (figure)
micro-moves, 130
mimicking in, 122–123, 126
not doing behavior, 120
objectives of, 119, 125
practice, 124, 126
problem, 119–124
rebranding, 125
studenting behaviors, 120–124
thinking classroom implementation, 124–126
trying it on their own behavior, 122–123
Try This activity, 131
worked solution, 128–129
workspace. *See* Workspace, student
Homogenous groups, 39
How to Solve It (Pólya), 19

answering questions, 95
collaborative groups, 54
consolidation, 183
evaluation, 226
formative assessment, 250
furniture arrangement, 78
grading, 276
hints and extensions, 166
homework, 130
note taking, 205
student autonomy, fostering, 141
task allocation, 115
tasks, 35
workspace, student, 67
Middle school
challenge and skill balance, 152–154
highly engaging thinking tasks in, 21
Mimicking, 9–10, 20
advantage, 30–31
disadvantage, 31
in homework, 122–123, 126
task, 26, 28–29
and task allocation, 114
Mobility of knowledge, 47–49
Mode of engagement, 158–160
Myth of objectivity, 255–256
The Myth of Objectivity (Romagnano), 255

Navigation instruments, 234–242, 245–249, 287–293
NCTM Principles and Standards, 19
Nexus of control, 43–45
Non-curricular tasks, 28–31, 152, 160
defined, 24
highly engaging, 31–32
students in flow on, 160 (figure)
Non–mathematics related activity, 104, 136
Non-permanent surfaces, 62 (figure)
Non-routine tasks, 20
Non-thinking behavior, 5
from now-you-try-one task, 8–10
Not doing homework behavior, 120
Note taking, 57, 187
accuracy, 203–204
chronological/spatial sequencing and, 189–190
in collective synergy to individual understanding transformation, 289–293
competencies, 199
consolidating from bottom and, 202
copying, 189, 191

dead notes, 188–191
FAQ, 201–204
fill-in-the-blank notes, 187, 191, 201
for future forgetful selves, 193–200, 204
graphic organizers for, 195–200, 204
hints for, 202–203
issue, 187
I-write-you-write notes, 187, 201
live notes, 188–191
macro-moves, 205
meaningful notes, 193–195, 202, 285–286
micro-moves, 205
mindful notes, 193–195
as non-thinking activity, 192
online, 204
problem, 187–192
purpose of, 187
student notes, 201 (figure)
thinking classroom implementation, 193–201, 285–286
Try This activity, 206–207
Type I graphic organizer, 195, 196 (figure)
Type II graphic organizer, 196, 196 (figure)
Type III graphic organizer, 197 (figure), 198
Type IV graphic organizer, 198, 199 (figure)
using examples and annotation, 194
value of, 192
worked examples, 198–200
Now-you-try-one task, 8–10
consolidation after, 171
studenting behaviors distribution on, 10, 10 (figure)
Numeracy tasks, 21–23

Objective grading, 255–256
Objectivity
myth of, 255–256
tyranny of, 257
Observable competency, 220, 224
Observational rubric, 214, 224
O'Connor, K., 254, 257–258
Off-task behavior, 9
Online
grading, 271–272
note taking, 204
portfolios, 274
On-task behavior, 9
Open ended tasks, 23–24

Studenting behavior, 7
  cheating, 121–122, 138
  defined, 7
  distribution of, 10, 10 (figure)
  faking, 9
  getting help, 122
  homework, 120–124
  learning and, 7–8
  mimicking, 9–10
  norms, 13
  not doing homework behavior, 120
  now-you-try-one task, 8–10
  slacking, 9
  stalling, 9
  trying it on their own, 10,
      122–124
Student-responsibility practices, 292
Students
  active and passive state, 103
  anonymous state, 61–62
  asynchronous activity, 145
  fixed mindset, 217
  growth mindset, 217
  non-thinking, 4–6, 6 (figure), 7–11
  notes, 201 (figure). *See also* Note
      taking
  socially defined roles, 85–86
  synchronous activity, 145
  teachers answering student questions.
      See Answering questions
  thinking. *See* Thinking
  in traditional classroom, 3 (figure)
  workspace. *See* Workspace, student
Subjective knowledge, 256
Summative evaluation, 231
See also Evaluation
Super organized classroom, 72
Symmetrical furniture placement, 74
Synchronous activity, 145
System theory, 283

Table arrangement. *See* Furniture
      arrangement
Task allocation, 99
  at beginning of lesson, 102–103
  classroom workspace and, 111
  of data analysis/graphing, 113
  at end of lesson, 101
  FAQ, 111–114
  forms, 104–110
  groundwork before, 106, 109
  issue, 99
  macro-moves, 115
  micro-moves, 115

at middle of lesson, 101
  mimicking and, 114
  in primary classrooms, 114
  problem, 99–100
  raising hands after, 135
  among sitting students, 103–104
  among standing students, 103–104,
      103 (figure), 111
  storytelling, 114
  tax collector task, 107–110
  from textbook/workbook, 100
  by textual instructions, 106–112
  thinking classroom implementation,
      101–110
  timing of, 101–103
  Try This activity, 116–117
  by verbal instructions, 105–112
  in worksheet form, 104–105
  writing on vertical surface, 105
Tasks, 19
  card tricks, 21–22
  curriculum, 26, 28–31, 33
  elementary classroom students in
      thinking, 26 (figure)
  FAQ, 30–35
  good problem-solving, 20
  high-ceiling, 23
  highly engaging thinking, 21–22
  issue, 19–23
  low-floor, 23
  macro-moves, 35
  micro-moves, 35
  mimicking, 28–29
  non-curricular, 28–31, 152, 160
  non-routine, 20
  now-you-try-one, 8–10
  numeracy, 21–23
  off-task behavior, 9
  on-task behavior, 9
  open ended, 23–24
  open-middle, 23–24
  problem, 24–25
  problem-solving, 19–20, 25
  resources, 99
  rich, 20, 22, 24–25
  scripted curriculum, 27–30
  ski trip fundraiser, 23
  textual, 106–112
  thick slicing sequence, 155
  thinking classroom implementation,
      25–30, 283–284
  thin slicing sequence, 155–156,
      155 (figure)
  true or false, 24

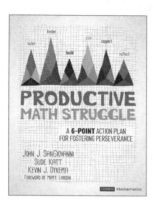

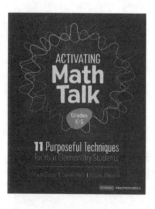

ALL students should have the opportunity to be successful in mathematics!

Trusted experts in mathematics education offer clear and practical guidance to help students move from surface to deep mathematical understanding, from procedural to conceptual learning, and from rote memorization to true comprehension. Through books, videos, consulting, and online tools, we offer a truly blended learning experience that helps you demystify mathematics for students.

**ROBERT Q. BERRY III,
BASIL M. CONWAY IV,
BRIAN R. LAWLER,
JOHN W. STALEY**

Teach mathematics through the lens of social justice to connect content to students' daily lives and fortify their mathematical understanding.

High School

**JOHN J. SANGIOVANNI,
ERIC MILOU**

Give math practice routines a makeover with these energizing warmups designed to jumpstart reasoning, reinforce learning, and instill math confidence in students!

Elementary, Middle School, High School

**BETH MCCORD KOBETT,
KAREN S. KARP**

Encourage productive struggle by identifying teacher and student strengths, designing strengths-based instruction, discovering students' points of power, and promoting strengths in the school community.

Grades K–6

**KAREN S. KARP, BARBARA J. DOUGHERTY,
SARAH B. BUSH**

Lead educators through the collaborative process of establishing a consistent learner-centered and equitable approach to mathematics instruction through team building.

Elementary, Middle School, High School

CORWIN Mathematics

A SAGE Publishing Company

Helping educators make the greatest impact

CORWIN HAS ONE MISSION: to enhance education through intentional professional learning.

We build long-term relationships with our authors, educators, clients, and associations who partner with us to develop and continuously improve the best evidence-based practices that establish and support lifelong learning.